Human Factors Analysis of 23 Cyberattacks

As cyber threat actors have become more sophisticated, data breaches, phishing attacks, and ransomware are increasing, and the global cybercrime damage in 2021 was $16.4 billion a day. While technical issue analyses are fundamental in understanding how to improve system security, analyzing the roles of human agents is crucial. *Human Factors Analysis of 23 Cyberattacks* addresses, through examples, the human factors behind cybersecurity attacks.

Focusing on human factors in individual attack cases, this book aims to understand the primary behaviors that might result in the success of attacks. Each chapter looks at a series of cases describing the nature of the attack through the reports and reviews of the experts, followed by the role and human factors analysis. It investigates where a human agent's intervention was a factor in starting, discovering, monitoring, or suffering from the attacks. Written in an easy-to-understand way and free from technical jargon, the reader will develop a thorough understanding of why cyberattacks occur and how they can be mitigated by comparison to the practical examples provided.

This title will appeal to students and practitioners in the fields of ergonomics, human factors, cybersecurity, computer engineering, industrial engineering, and computer science.

Abbas Moallem, Ph.D., is a consultant and adjunct professor at San Jose State University, California, where he teaches human–computer interaction, cybersecurity, information visualization, and human factors. He is also the program chair of HCI-CPT, the International Conference on HCI for Cybersecurity, Privacy, and Trust, and the track chair of the International Conference on Human Factors in Cybersecurity.

Dr. Moallem is the editor of *Human-Computer Interaction and Cybersecurity Handbook*, *Smart and Intelligent Systems*, *The Human Elements in Artificial Intelligence, Robotics, and Cybersecurity*, *Cybersecurity Awareness Among Students and Faculty*, *Understanding Cybersecurity Technologies*, *A Guide to Selecting the Right Cybersecurity Tools*. He is also the editor of a book series from CRC Press titled *The Human Element in Smart and Intelligent Systems.*

The Human Element in Smart and Intelligent Systems

Series Editor: Abbas Moallem

This series aims to cover the role of the human element in all aspects of smart and intelligent systems. It will include a broad range of reference works, textbooks, and handbooks. The series will look for single-authored works and edited collections that include, but not limited to, automated driving, smart networks and devices, cybersecurity, data visualization and analysis, social networking, smart cities, smart manufacturing, trust and privacy, artificial intelligence, cognitive intelligence, pattern recognition, computational intelligence, and robotics. Both introductory and advanced materials for students and professionals will be included.

Smart and Intelligent Systems: *The Human Elements in Artificial Intelligence, Robotics, and Cybersecurity*
Abbas Moallem

Understanding Cybersecurity Technologies: *A Guide to Select Right Cybersecurity Tools*
Abbas Moallem

Human Factors Analysis of 23 Cyberattacks
Abbas Moallem

For more information on this series, please visit: https://www.routledge.com/The-Human-Element-in-Smart-and-Intelligent-Systems/book-series/CRCHESIS

Human Factors Analysis of 23 Cyberattacks

Abbas Moallem

CRC Press
Taylor & Francis Group
Boca Raton London New York

CRC Press is an imprint of the
Taylor & Francis Group, an **informa** business

Front cover image: Luyali/Getty Images.

First edition published 2025
by CRC Press
2385 NW Executive Center Drive, Suite 320, Boca Raton FL 33431

and by CRC Press
4 Park Square, Milton Park, Abingdon, Oxon, OX14 4RN

CRC Press is an imprint of Taylor & Francis Group, LLC

ISBN: 978-1-032-60862-4 (hbk)
ISBN: 978-1-032-61051-1 (pbk)
ISBN: 978-1-003-46173-9 (ebk)

DOI: 10.1201/9781003461739

Typeset in Times
by codeMantra

Contents

PART TWO *Cyberwar and Cyberespionage*

PART FIVE Data Privacy

PART SIX Trust

PART EIGHT Blackmail—Hacktivism

PART NINE Conclusion

Acknowledgments

I am grateful to Professor Wojciech Cellary, University WSB Merito in Poznan, and Dr. Francisco Corella, CEO at Pomcor, who reviewed drafts of this book and provided valuable comments and suggestions. Without their support, this book would not have become a reality.

Introduction

Every day, cyberattacks are increasingly in the daily news. The attacks are growing in number and getting worse each year. Threat actors have become more sophisticated, resulting in increasing numbers of data breaches, phishing attacks, and ransomware. As a result, the cost of cyberattacks has also constantly increased. The global cybercrime damage in 2021 was $16.4 billion a day, $684.9 million an hour, $11 million per minute, and $190,000 per second. It is estimated that global cybercrime costs will grow by 15% per year over the next five years, reaching an annual cost of $10.5 trillion worldwide by 2025—up from $3 trillion in 2015 [1].

The press routinely reports cyberattacks. According to some reports, in 2018, there were 80,000 cyberattacks happening per day [2]. That amounts to over 30 million attacks occurring in a year. However, there are no reliable statistics about how many cases happen in each country. The reported cases tend to be the ones that affect a considerable population or significant companies, or they are reported by cybersecurity monitoring organizations.

Many cyberattack cases are not reported at all. This is especially true of cases conducted by national security agencies to collect data from their adversaries or committed as part of a larger cyberwar among different nations. There are also many cyberattacks done by investigative organizations, groups, or individuals. These attacks may be done to access companies' intellectual property, collect information about other countries, or commit organized crimes. The victim organization may not report these cases if no users are involved, in order to protect its reputation.

Regardless, every day many cyberattack cases are revealed. Each of these cases is indicative of certain types of cyberattacks. In general, the attacks that become public knowledge are analyzed by cybersecurity experts and different companies providing cybersecurity solutions to help encourage others to allow them to protect their systems. All the experts who investigate cases aim to understand the technical vulnerability of cyberattack victims. In addition, the motivations of the attackers, the data compromised, and how compromised are generally the focus of attention.

After reviewing many cyberattack cases published in the press, I noticed that most of the analyses focus on the technical vulnerability of the systems, technologies, or software used, with few focusing on the roles of human agents. However, human agents are essential and are important elements at many different levels. Human agents on the attacker side include the people who decide to create the attacks and the behavior of Information Technology (IT) agents on the attackers' side. On the other side, the behavior of agents on the defender side includes the victims of the attack, managers of companies, and human agents who build the networks or code the software.

Analyzing the human agent component of cyber-security is a crucial element and an addition to existing technical issues analyses. Even when all security measures are taken, human error or behavior can still make all systems vulnerable [3].

This book intends to focus on the human factors in each attack case. The aim is to understand the primary behaviors that might result in the success of attacks.

I have selected 23 major cyberattack cases based on their characteristics representing the different types of attacks. Each chapter looks at one case and describes it, as well as what we know about it through the reports and reviews of the experts, followed by role and human factors analysis. I also investigate if the human agent's intervention played a role in starting, discovering, monitoring, or suffering from the attacks.

The description of each case is simplified to facilitate understanding for readers who might not be familiar with technical terminologies. Additionally, the use of technical jargon is limited. Thus, the focus can be placed on human factors in each case.

The first three chapters define and review attackers, hackers' motivations, and humans' roles at each phase. In the following 24 chapters, I will review each famous case, and in the final chapter, I will summarize factors shaping vulnerable behaviors.

Abbas Moallem

REFERENCES

1. Morgan S. (2020). "Cybercrime to Cost the World $10.5 Trillion Annually by 2025," Cybercrime Magazine, November 13, 2020. https://cybersecurityventures.com/cybercrime-damage-costs-10-trillion-by-2025/.
2. Firch J. (2021). "Cyber Security Statistics, The Ultimate List of Stats, Data & Trends," Purplesec, Retrieved April 21, 2022. https://purplesec.us/resources/cyber-security-statistics/#:~:text=Cyber%20Security%20Risks,30%20million%20attacks%20per%20year.
3. Kaspersky. (2022). "The Human Factor in IT Security: How Employees Are Making Businesses Vulnerable from Within," Kaspersky Blog. https://www.kaspersky.com/blog/the-human-factor-in-it-security/ (accessed September 20, 2022).

Part One

Human and Motivation

- Human Agents
- Cyberattackers versus Hackers
- Cyberattackers' Motivations

1 Human Agents

Abstract

This chapter will describe human agents' roles in cyberattacks on both the attack and defense sides.

Humans are essential components of any system. They decide to build systems and then manage, monitor, and maintain them before finally using them. In cybersecurity, humans move through these same actions. They design software and provide the intelligence to build it out. Humans are the starting point for creating and coding the applications that people and enterprises use, which also means humans are behind the software built to attack.

Consequently, human agents must provide programming and build intelligent systems. Without humans, no computer systems, robots, or smart and intelligent systems exist. We can better understand human agents' roles in cybersecurity by identifying whether a person is part of one of three groups: attackers, defenders, or victims (Figure 1.1).

The cyberattacker and defender groups can be broken down further into distinct human agent roles.

To understand the human agent's vulnerability and help protect the system, it is necessary to begin by understanding the behavior of humans in each role. This chapter will describe these roles from a cybersecurity standpoint.

HUMAN ROLES ON THE ATTACK SIDE

In the cyberattacker group, there are four general roles. First, humans are the decision-makers in attacking systems. Second, the architects who design the attacks. Third, IT agents who execute the attack. Fourth, agents who monitor and control the outcome of the attacks (Figure 1.2).

Attack Decision-Makers

Attack decision-makers are the people who decide to attack a system. These agents might be organizations, groups, institutions, state agencies, criminal organizations, or private enterprises. When discussing cyberattacks, people often think criminal organizations are the attack decision-makers. This belief is supported by the fact that most cyberattack cases that come to public knowledge are attacks that target individuals for financial crimes or are revealed by the investigation. However, these attacks might not be the foremost part of cyberattacks as many cyberattacks are not readily shared with the public. For example, an intelligence agency of a country might want to break into another country's infrastructure systems, hack into its leader's accounts, or sabotage its manufacturing system.

DOI: 10.1201/9781003461739-2

FIGURE 1.1 Human roles in cyberattackes: attack, defence, and application technologies.

Understanding the attackers' motivations is fundamental for a good analysis of cyberattacks from a human factors point of view. Motivation provides insights into why individuals commit cyberattacks. It helps to understand the underlying factors that drive cyberattackers' behavior, whether it's financial gain, revenge, ideology, or other personal or social reasons. For example, the resources from the number of human agents to infrastructure and money that a government agency might put into attacks might not be comparable to a private investigation or even a prominent

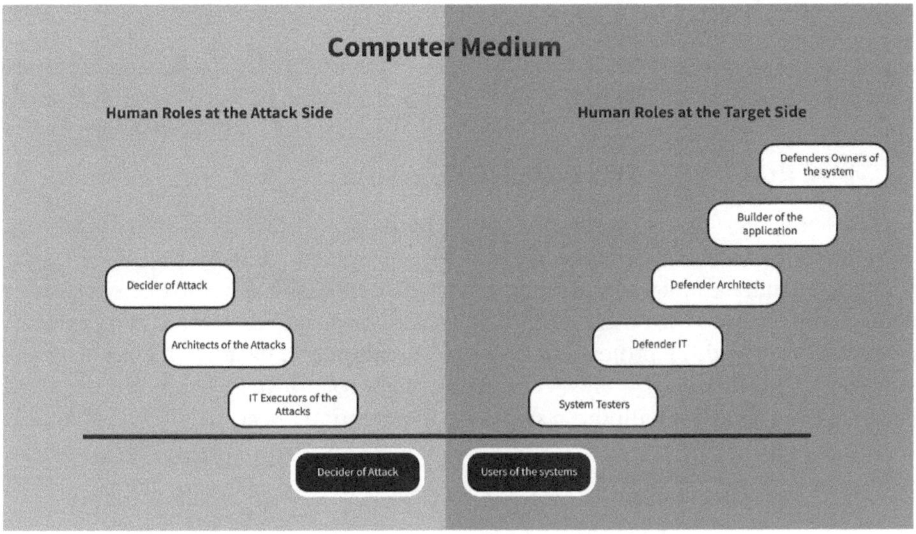

FIGURE 1.2 Human agents' roles at attack and target sides.

criminal organization. On the other hand, the attackers' motivation(s) might explain what they might do with the data they acquired from the attacks. For example, the Stuxnet attack on the Iranian nuclear facilities was impossible without considerable resources being allocated by a state involved in this case, such as the United States and Israel (see Chapter 7).

Attack Architects

Attack architects are the people with the expertise to organize the attack. Cyberattacks that require massive resources might not be built like a small criminal organization, or an individual might design an attack. In other words, different expertise will be needed based on the size and scale of the intended attack, as well as the motivation and resources allocated. For instance, attack architects may use the expertise of behavioral psychologists and human factors experts to predict the behaviors of the defenders to their attack to construct a stronger attack. For example, it is hard to believe that the SolarWinds attack was possible without tremendous infrastructure and human resources (Chapter 4).

Attack Executors

Attack executors are the IT professionals who carry out the decided and designed attack. The precise execution of the attack and human agents' vigilance are fundamental to the attacker's success. For example, any flaws in the execution of an attack might reveal the attacker's identity. The humans in these roles can be considered a point of vulnerability—human agents who have the potential to become whistleblowers, for instance.

Agents Monitoring Attacks

Agents who monitor and control the outcome of the attacks, or hackers and ransomware designers who sell the ransomware for financial gain [2,3].

HUMAN ROLES ON THE DEFENDERS' SIDE

The defender group is composed of several key roles. First, there are the decision-makers who determine which defense systems should be implemented. Second, defender architects and defense management teams develop and oversee the applications that run the business, ensuring that systems are monitored and protected from attacks. Third, IT professionals design, monitor, and evaluate the security and vulnerabilities of these systems, playing a crucial role in their defense. Lastly, the end-users who interact with these systems can potentially become victims of attack.

Defense Decision-Makers

The attack target might be companies, organizations, state agencies, or criminal organizations. The same entities, which may be the attack decision-makers, are also the ones who are attack targets. As a defender, however, their objectives are reversed. A criminal organization might be attacked by law enforcement. A company might be attacked by its adversaries for competitive advantages, money, or to cause reputational harm. The behaviors of the owners of the systems and management teams, in this case, are fundamental. Since they allocate the resources to protect the entities' digital assets, operation, reputation, and even existence (Example of Nortel Insider Treat) [4].

Defender Architects

The behavior, expertise, and knowledge of the system architects play a role in building a secure system against cyberattacks [1].

Defender architect builds the application. Application coding requires consideration of security issues. With probably billions or trillions of lines of code written, building and selecting defensive applications is a challenge in terms of security. Are all applications built with security in mind? Was an application built 20 years ago made with security standards? Consequently, the human agents and process of building codes are part of all applications' security challenges and vulnerabilities.

Defender IT and System Testers

IT professionals who monitor, manage, and evaluate the security and vulnerability of the system are fundamental from a human factors point of view. There are several famous cases of the errors of the IT managers that have created vulnerability of the system to attacks (TJ Max, Target, SolarWinds Cases—Chapters 4, 23, and 24).

The agents testing the systems (penetration testers) also protect the systems from attacks. The tester's expertise, the extent of testing, and creativity in checking systems' vulnerabilities are paramount. After all, the attackers are sometimes more imaginative in discovering the vulnerability of the systems than testers.

Users of the Systems

Ultimately, humans are also victims of attacks. Employees within an organization can be particularly vulnerable. With potentially hundreds of thousands of employees, user behavior can create significant vulnerabilities, ranging from falling victim to phishing schemes and insider threats to being manipulated through social engineering—every cybersecurity expert's worst nightmare. There are countless examples where the actions of a single employee have led to vulnerabilities, making attacks successful (Target case, Chapter 23).

VICTIMS OF CYBERATTACKS

Millions of people are subject to cyberattacks every year, and they are constantly on the rise. While there is overlap between defenders and victims, the undefended victim, especially in the context of the individual, is also a type of human agent whose behavior must be considered. How do victims and potential victims behave when protecting their digital assets and computer systems? What does a cyberattack victim do before, during, and after an attack? What risks are potential victims aware of? How does a victim of a cyberattack change their behavior? What lessons do they learn? Does their behavior change, and if so, how? These questions need to be answered to create tools, awareness, and solutions to protect people from cyberattacks.

REFERENCES

1. Miller J. (2021). "The importance of Humans in Cybersecurity," bitlyft.com, May 20, 2021. https://www.bitlyft.com/resources/importance-of-humans-in-cybersecurity.
2. United State Attorney's Office. (2022). "Hacker and Ransomware Designer Charged for Use and Sale of Ransomware, and Profit Sharing Arrangements with Cybercriminals," www.justice.gov, May 16, 2022. https://www.justice.gov/usao-edny/pr/hacker-and-ransomware-designer-charged-use-and-sale-ransomware-and-profit-sharing.
3. U.S. Department of the Treasury. (2024). "United States Sanctions Senior Leader of the LockBit Ransomware Group," May 7, 2024. https://home.treasury.gov/news/press-releases/jy2326.
4. Naraine R. (2012). "Nortel Hacking Attack Went Unnoticed for Almost 10 Years," ZD Net, February 14, 2012. https://www.zdnet.com/article/nortel-hacking-attack-went-unnoticed-for-almost-10-years/#google_vignette.

2 Cyberattackers versus Hackers

Abstract

This chapter defines the terms "cyberattacker" and "hackers" and reviews the evolution of hacker classification over time.

INTRODUCTION

Cyberattackers are people, organizations, or groups that deliberately attempt to breach the information system(s) of individual people, organizations, enterprises, and different government institutions. In addition, cyberattackers seek some benefit from collecting data, monitoring activities, and disrupting the victim's network.

Hackers, in general, are information technology (IT) professionals who use their technical knowledge to achieve a goal or overcome an obstacle within a computerized system by non-standard means. IT professionals generally work for different organizations (criminals or legitimate states), although they occasionally operate independently. Hackers might break into a system and access data otherwise unavailable to them. Hackers do not always violate the law; there are situations where hackers can break into a system legally. For example, law enforcement agencies sometimes use hacking techniques to collect evidence on criminals and other malicious actors. Hackers might also be hired to try to hack a system (known as "penetration tests") to find out and report the system's vulnerability to help the system operator remove that vulnerability.

Cyberattackers are commonly called *hackers*. However, hackers and cyberattackers are not necessarily interchangeable terms with the same meaning. The term hacker is broad because it describes the activity [1]. It is also commonly used to describe individuals who engage in unauthorized activities, such as exploiting vulnerabilities in computer systems for malicious purposes. Hackers might have a broader connotation, both positive and negative associations, depending on the context in which it's used. Hacker classification has also evolved through time.

Earlier studies categorized hackers into seven groups based on their technical ability from lowest to highest as follows [2]:

- Tool kit/Newbies
- Cyberpunks
- Internals
- Coders
- Old-guard hackers
- Professional criminals
- Cyberterrorists

 DOI: 10.1201/9781003461739-3

Another classification of hackers categorized them into the following groups by their skills or personas [3,4]:

- Novices
- Cyberpunks
- Insiders
- Old Guards
- Professionals
- Hacktivists
- Nation States
- Students
- Petty Thieves
- Digital Pirates
- Online Sex Offenders
- Crowdsourcers
- Crime Facilitators

In another classification, hackers were classified based on their knowledge exchange behaviors through posting behavior and content analysis. Then, statistical learning theory was used to classify them into four groups:

- Guru hackers
- Casual hackers
- Learning hackers
- Novice hackers

This type of classification does not reflect the motivations of the hackers. However, it might be a good indication of how sophisticated their operation might be and the support that they might have [3].

In another commonly used classification, hackers are classified based on their intentions and methods. The "hat" system is a popular way to describe different types of hackers, which is broken down below (Table 2.1).

White Hat Hackers: White hat hackers are generally hired to test systems and networks by trying to break into them to improve the system's cybersecurity. In this case, the hackers are usually authorized by an organization. The most common example of this type of hacker is penetration testers who try to break into a company's computer systems and networks. These authorized tests (Penetration Testers) help identify security vulnerabilities and weaknesses before malicious hackers can exploit them. Many agencies and individual consultants specialize in this type of testing.

Black Hat Hackers: Black hat hackers are those who violate computer security for their own personal or organizational profit. Black hat hackers are motivated by self-serving reasons, such as financial gain, revenge, or spreading havoc. With the number of cyberattacks increasing, sophisticated gangs are

TABLE 2.1
The Color of "Hat" Used to Classify the Type of Hackers

Hacker Types	Description
White Hat	Hired to test systems and networks by trying to break into them to improve the cybersecurity of the system
Black Hat	Violates computer security for their own personal or organizational profit
Red Hat	Activists who try to attack the black hat hackers by using legal or illegal ways to achieve their goals
Gray Hat	They do not have malicious intent like black hat hackers, but they might violate the privacy and numerous other laws of the country where they are located. We can also consider hacktivists or whistleblowers gray hat hackers
Green Hat	They may not intentionally want to cause harm to others but may do so while practicing their skills
Blue Hat	They can be former employees of a company, institution, or government

operating worldwide. Some of these organized crime gangs are well-known, such as the Cobalt Cybercrime Gang, behind the Carbanak and Cobalt malware attacks that targeted 100 financial institutions worldwide in more than 40 countries. Another famous one is the Lazarus Gang, which is believed to be behind the Sony Pictures leak, which also stole terabytes of confidential data, deleted some files and threatened to leak the information if Sony refused the hackers' demands [5].

Gray Hat Hackers: Gray hat hackers do not have malicious intent like black hat hackers. However, they might violate the privacy and numerous other laws of the country where they are located. Gray hat hackers might interrupt a system to illustrate their points or do this out of curiosity, rather than for personal profit or gain.

Gray hat hacking is illegal, as the hacker has not received permission from an organization to attempt to infiltrate their systems. We can also consider hacktivist groups, which can be individual or anonymous hackers, whistleblowers, or gray hat hackers. For example, Anonymous is a hacktivist group known for its various cyberattacks against several government institutions, agencies, corporations, and the Church of Scientology [6].

Another example would be Edward Joseph Snowden, a former computer intelligence consultant who leaked highly classified information from the National Security Agency (NSA) in 2013 [7,8].

Red Hat Hackers: Red hat hackers are activists who try to attack systems by deploying cyberattacks against them using illegal means to achieve their goals. Red hat hackers might use different ways to achieve their goal, such as infecting systems with malware or launching denial-of-service (DDoS) attacks. DoS floods the target site with many requests, consuming its capacity and rendering it unable to respond to legitimate requests.

Sometimes, Red hat hackers' motivation might be ideological, such as targeting people they strongly disagree with. For example, these hackers might attack black hackers, but their actions would still be illegal. There are several known red hat hackers; one is the Anonymous group that hacked the Church of Scientology [9,10].

Green Hat Hackers: Green hat hackers are novice hackers who may not intentionally want to cause harm to others but do so while practicing their skills. We can call them hackers in training. These types of hackers are not very familiar with laws and cybersecurity techniques. They can also include hackers who commit these acts to get attention from their peers, or juveniles who don't have complete knowledge of hacking and are taking part out of curiosity.

An example would be the teenager in Lodz, Poland who hacked the city's tram system with a homemade transmitter that tripped rail switches and redirected trains, a prank that derailed four trams and injured a dozen people [11].

Blue Hat Hackers: Blue hat hackers take personal revenge. They can be former employees of a company, institution, or government. Blue hat hackers use malware and deploy various cyberattacks on their enemies' servers/networks to harm their data, websites, or devices.

Blue hat hackers might use various hacking techniques to bypass authentication mechanisms to gain unauthorized access to their targets' email clients or social media profiles. They do so to damage a company's reputation, steal its customers' confidential data, release it to the public, and damage a company or organization's reputation.

Some organizations use blue hat hackers to test new software and find security vulnerabilities before a software release. Sometimes, companies organize periodic conferences for blue hat hackers to find bugs in their crucial online systems.

CYBERATTACKERS

Most reports use the term hacker, which generally refers to someone who launches deliberate and malicious attacks on computer systems, networks, or digital infrastructure. Cyberattackers may use various techniques, such as malware, phishing, social engineering, or exploitation of software vulnerabilities, to gain unauthorized access, steal data, disrupt operations, or cause other harm. The term "cyberattackers" is generally used without specifying the category or mentioning that these can include criminal organizations, nation-states, spy agencies, or law enforcement agencies.

For the purpose of this book, we will use "cyberattackers," which will cover all legitimate or non-legitimate hacking categories. All will be considered an attack on systems that victims cannot enter, view, use, or modify, and will include the use of anonymity tools (such as a VPN or the Dark Web) to mask the attackers' identities online. Cyberattacks might also be used illegally by law enforcement and security agencies and employed by nation-state actors as legal and illegal warfare weapons.

This chapter reviews the specificity of cyberattackers and hackers from an operational point of view based on their motivations.

CONCLUSION

Overall, not all types of hackers—such as those with good intentions, criminal intent, or state spy and security agencies—follow the same motivation and aims. For example, how would we classify a country's spying agency that hacks another country's system? Are the hacking agents wearing white hats, black hats, gray hats, etc.? Let's consider the example of Stuxnet, a computer worm that targeted Iran's nuclear facilities and was used to attack Iranian nuclear facilities [12]. Was it legal or illegal? Was the hacker a white hat or a black hat? How would one classify Edward Joseph Snowden? Consequently, it can be challenging to organize hackers. Thus, for the sake of simplicity, we will use the term "cyberattackers" rather than hackers in this book.

REFERENCES

1. Post J. (1996). "The Dangerous Information System Insider: Psychological Perspectives." https://www.infowar.com/.
2. Rogers M. (1999). The psychology of hackers: A new taxonomy. *RSA World Security Conference,* San Jose, CA.
3. Zhang Z., Tsang A. et al. (2015). The Classification of Hackers by Knowledge Exchange Behaviors. *Information Systems Frontiers.* Vol. 17, pp. 1239–1251. https://pweb.fbe.hku.hk/~mchau/papers/HackerForum_ISF2015.pdf.
4. Chng S., Han Y. L., Kumar A., Yau D. (2022). Hacker Types, Motivations, and Strategies: A Comprehensive Framework. *Computers in Human Behavior Reports.* Vol. 5, p. 100167. https://doaj.org/article/20dad15a8693478e9b765bd9c7d78d91.
5. Van Beveren J. (2000). A Conceptual Model of Hacker Development and Motivation. *Journal of E-Business.* Vol. 1, No. 2, pp. 1–9. https://citeseerx.ist.psu.edu/viewdoc/download?doi=10.1.1.586.3584&rep=rep1&type=pdf.
6. Balita-Centeno L. (2020). "The 5 Most Notorious Organized Cybercrime Gangs," Makeuseof.com, November 3, 2020. https://www.makeuseof.com/notorious-cyber-crime-gangs/.
7. Vamosi R. (2008). "Anonymous Hackers Take on the Church of Scientology," CNET, January 25, 2008. https://www.cnet.com/news/privacy/anonymous-hackers-take-on-the-church-of-scientology/.
8. Greenwald G., MacAskill E., Poitras L. (2013). "Edward Snowden: The Whistleblower behind the NSA Surveillance Revelations," The Guardian, June 11, 2013. https://www.theguardian.com/world/2013/jun/09/edward-snowden-nsa-whistleblower-surveillance.
9. Crimesider Staff. (2010). "Scientology Hacked; Church Website Cyber-Attacker Gets Year in Prison for "Anonymous" Invasion," CBS News, May 27, 2010. https://www.cbsnews.com/news/scientology-hacked-church-website-cyber-attacker-gets-year-in-prison-for-anonymous-invasion/#textLOS20ANGELES2028AP2920A20Nebraska20man20has20beensupervised20release20after20he20gets20out20of20prison.

10. Stouffer C. (2021). "An Overview Plus High-Profile Groups and Example," Norton Life Lock, September 8, 2021. https://us.norton.com/internetsecurity-emerging-threats-hacktivism.html.
11. Squatriglia C. (2008). "Polish Teen Hacks His City's Trams, Chaos Ensues," WIRED, January 11, 2008. https://www.wired.com/2008/01/polish-teen-hac/.
12. Fruhllinger J. (2022). "Stuxnet explained: The first known cyberweapon CSO 8(US), August 31 2022. https://www.csoonline.com/article/562691/stuxnet-explained-the-first-known-cyberweapon.html.

3 Cyberattackers' Motivations

Abstract

This chapter reviews different cyberattackers' main motivations and incentives, and various theoretical frameworks used to analyze attackers' motivations in cybersecurity.

INTRODUCTION

A motive is usually used to explain why a person commits a crime. For example, one of the primary acts of a detective investigating a murder case is determining who would benefit from the murder since the motive is the underlying reason for a crime.

In cybersecurity, the motivation of cyberattackers (criminal individuals, organizations, government spying agencies, or companies) is also fundamental to understanding why a cyberattack happens and what the attackers intend to do with what they get; in other words, what would be their incentives? To better grasp these concepts, we examine the motivations of cyberattackers who unlawfully access an organization's digital assets, as well as companies or legal entities that gather information from people without their consent, thereby violating their privacy when using their services. This is distinct from white hat hackers, who are hired to test systems and networks by attempting to breach them to enhance cybersecurity.

This chapter reviews the main motivations and incentives of different cyberattackers (Figure 3.1).

FINANCIAL GAIN

One of the main motivations behind many cyberattacks is financial gain, whether through fraudulency, getting money or information about people or organizations, getting their money from their bank accounts or credit cards, or through ransomware attacks on individuals or organizations, or different phishing attacks—fake eCommerce sites, phishing voice calls to people, etc.

Another way to obtain financial gain is to sell the information to interested parties through the Dark Web marketplace or to include identity theft. According to the U.S. Federal Trade Commission, 4.8 million identity theft and fraud cases were reported in 2020, up 45% from 3.3 million in 2019, primarily due to a 113% increase in identity theft complaints [1].

In recent years, ransomware attacks have been very profitable. According to the National Credit Union Administration [2], ransomware remains a growing cyber threat to smaller financial institutions, including credit unions.

DOI: 10.1201/9781003461739-4

FIGURE 3.1 Main motivations and incentives of different cyberattacks.

Many examples of financial fraud are reported daily (Part 3). These cases are widespread, including some well-known instances.

SABOTAGE, CYBERWAR, AND INTELLIGENT GATHERING

Each country's spies and security agencies are the other primary sources of cyberattacks. The cyberattackers' motivation varies based on each country's objective, resources, and specific needs. However, the primary reasons are accessing other countries' classified information, cyberwar attacks, and sabotage. One of the most famous cases is the SolarWinds case (Chapter 4).

Another famous case is Stuxnet, the sabotage of Iranian nuclear facilities. Stuxnet is a malicious computer worm that is believed to be used by Israel to cause substantial damage to the nuclear program in Iran [3] (Chapter 7).

Another case example would be the assassination of Qasem Soleimani, an Iranian general, through location tracking. Soleimani was assassinated as ordered by U.S. President Donald Trump in a targeted American drone strike on January 3, 2020, in Baghdad, Iraq [4].

FAKE NEWS SHAPING PEOPLE'S OPINIONS

Another motivation for cyberattacks is shaping people's opinions and influencing elections through fake news or misinformation of a target population. Cambridge Analytica is one such case (Chapter 10). Cambridge Analytica was a company that infamously attempted to use the psychological profiles of roughly 90 million

Facebook users to influence the exit of the United Kingdom from the European Union (known as the Brexit campaign) and Donald Trump's 2016 presidential campaign [5]. However, one would consider these types of cases to be more based on the sensitivity and secrecy of each case.

BLACKMAIL

Many private investigation companies are active in cyberspace by providing cyber investigations nationally and internationally. Cyber-security private investigation firms offer services and individual assistance to countries and companies. Depending on where they are located, they can use legal or illegal means using various tools that hackers generally use for their private services to gather information from social media platforms. Many also have experience using the Dark Web to track down hard-to-find data of people. Then, the information collected can be used for blackmailing, court cases, etc.

One of the most notorious cases is the News of the World phone-hacking scandal in the United Kingdom. Private investigators were involved in illegally obtaining information, which they provided to the News of the World, along with engaging in other unlawful activities. Jonathan Rees and his partner Sid Fillery, a former police officer, purchased information from former and serving police officers, customs officers, V.A.T. (Value Added Tax) inspectors, bank employees, burglars, and bloggers. Then, they would telephone the Inland Revenue, the DVLA, banks, and phone companies and deceive them into releasing confidential information. Rees then sold the information to the News of the World, the Daily Mirror, the Sunday Mirror, and The Sunday Times [6].

Another famous blackmail case is the Ashley Madison case where attackers copied the personal information of the site's user base and threatened to release their identifying information if Ashley Madison did not shut down immediately (Chapter 23).

SOLVING CRIME CASES

Different law enforcement agencies in countries might also use cyberattacks with and without warrants, depending on where they operate and the laws and regulations of each country.

According to Justsecurity.org, "Law enforcement officials have begun using commercial and bespoke hacking tools to interfere with computer systems to access and gather sensitive information, including individuals' locations, internet activities, communications, and personal files. At the same time, some of these hacking tools are developed in-house, and in many instances, the U.S. government has purchased these technologies from private companies" [7,8].

In another controversial case, a German state bought secret Swiss bank data to clamp down on tax evaders. According to reports, it is believed that German state of North Rhine-Westphalia have paid more than €5m for the data, which is the highest amount handed over for such information [8a].

ILLEGAL, IMMORAL, ILLICIT, DENUNCIATIONS, AND REVELATIONS

Whistleblowers expose information or activity within a private, public, or government organization. Their motivation is to inform the public to stop what they consider illegal, harmful actions of companies, organizations, or government agencies. Many whistleblowers might see some activities revealed through different means. In recent history, many whistleblower cases were the sources of revelation. The most famous such case is Edward Snowden [9].

However, many other cases reported revealed companies' and governments' activities. The case of a U.S. Army intelligence analyst who released the most extensive set of classified documents ever, published mainly by WikiLeaks, is another famous one [10].

There is also the case of James Glenn, a 42-year-old security specialist who revealed Cisco's serious security flaws and lost his job [11].

We can also add the hacktivist groups that hack into unauthorized networks to expose a believed injustice in this category. Their motivation is not necessarily to cause malicious harm but to draw attention to an important cause of the hacktivist group. Thus, the reason might be denouncing social injustice from the government, human rights, religion, or freedom.

One of the most famous hacktivist groups is Anonymous. Anonymous is a decentralized hacking group often launching against political figures, foreign governments, and religious groups [12]. Their targets include governments, multinational corporations, and influential individuals. Politically motivated hacktivism seeks to promote or upend a political agenda, sometimes to the extent of anarchy [13].

REVENGE

Revenge behavior is universal in humans. It is an emotional feeling that can come up in every human being. The revenge action causes unpleasant or painful hurt or harm to people, organizations, companies, or countries. Revenge behavior is broadly analyzed in social psychology. While revenge is an emotional behavior in all people, it may differ among people and across cultures.

Some people have more vengeful tendencies than letting an issue go. Revenge is also associated with pleasure [14]. It seems that some people are more vindictive than others. Some research suggests that vengeful people are motivated by power, authority, and a desire for status [15].

In cybersecurity, revenge behavior is mainly observed in insider threat cases. It appears that former employees are motivated, at least partly, by a desire to seek revenge [16]. Studies suggest that 84% of the incidents of inside threats were motivated at least partly by a desire to seek revenge [17]. It is interesting to observe that not only people but also organizations, groups, or countries might cyberattack to get revenge.

HAVING FUN AND CURIOSITY

Sometimes, the motivation of cyberattackers seems to be simply having fun. A famous hacker, Kevin David Mitnick, a convicted hacker who broke into 40 major corporations and then became a security consultant, once said, "The motivation for me was that it was fun and an intellectual challenge…" [18]. Another suspected hacker behind one of the biggest cryptocurrency heists has revealed that he did it "for fun." Over $600 million in cryptocurrency assets was stolen from Poly Network, a decentralized finance (Defi) platform. Poly Network said the hacker returned $342 million worth of crypto to the platform by August 12 [19]. In another case in 2013, Burger King's Twitter feed was hacked by hackers affiliated with the Anonymous collective, who quickly took it over and rebranded it with the rival McDonald's logo and name, and began using it to tweet McDonald's special offers [10,20].

TERRORIST ORGANIZATIONS

Terrorist organizations operating in a clandestine world might also cyberattack to harm their targets, sabotage critical infrastructures, or kill. The motivation of cyberterrorists is any terrorist acts to create death and destruction and consequently create fear. There are cyberattacks carried out by terrorists and terrorist groups [21].

In December 2015, the world witnessed the first known power outage caused by a malicious cyberattack. For example, three utility companies in Ukraine were hit by BlackEnergy malware, leaving hundreds of thousands of homes without electricity for 6 hours [22]. In another case, the Shadow Brokers hacked the U.S. National Security Agency (NSA) and were used in a computer cyberattack in more than 150 countries [23].

PRINCIPAL THEORETICAL FRAMEWORKS TO ANALYZE CYBERATTACK

Examining why people commit a crime is very important. Many theories have emerged over the years that can be used to analyze cyberattackers' motivations. These theories and frameworks continue to be explored, individually and in combination, as criminologists seek the best solutions to reduce crime types and levels. Then, various theoretical frameworks are used to analyze the motivations of different categories of attackers in cybersecurity (Figure 3.2).

RATIONAL CHOICE THEORY

Adam Smith's rational choice theory on people's economic and social behavior suggests that people perform a cost–benefit analysis to determine whether an option is correct. The extrapolation of this theory in cybersecurity generally postulates that people act in their self-interest and decide to commit a crime after weighing the risks (including getting caught and punished) against the rewards. Thus, a cybercriminal chooses to commit a cybercrime to maximize satisfaction by selecting the less punitive action that brings more benefits. Rational choice explicitly assumes a rational offender and emphasizes the criminal's advantage and motivation, which synchronize with cybercrime [24,25].

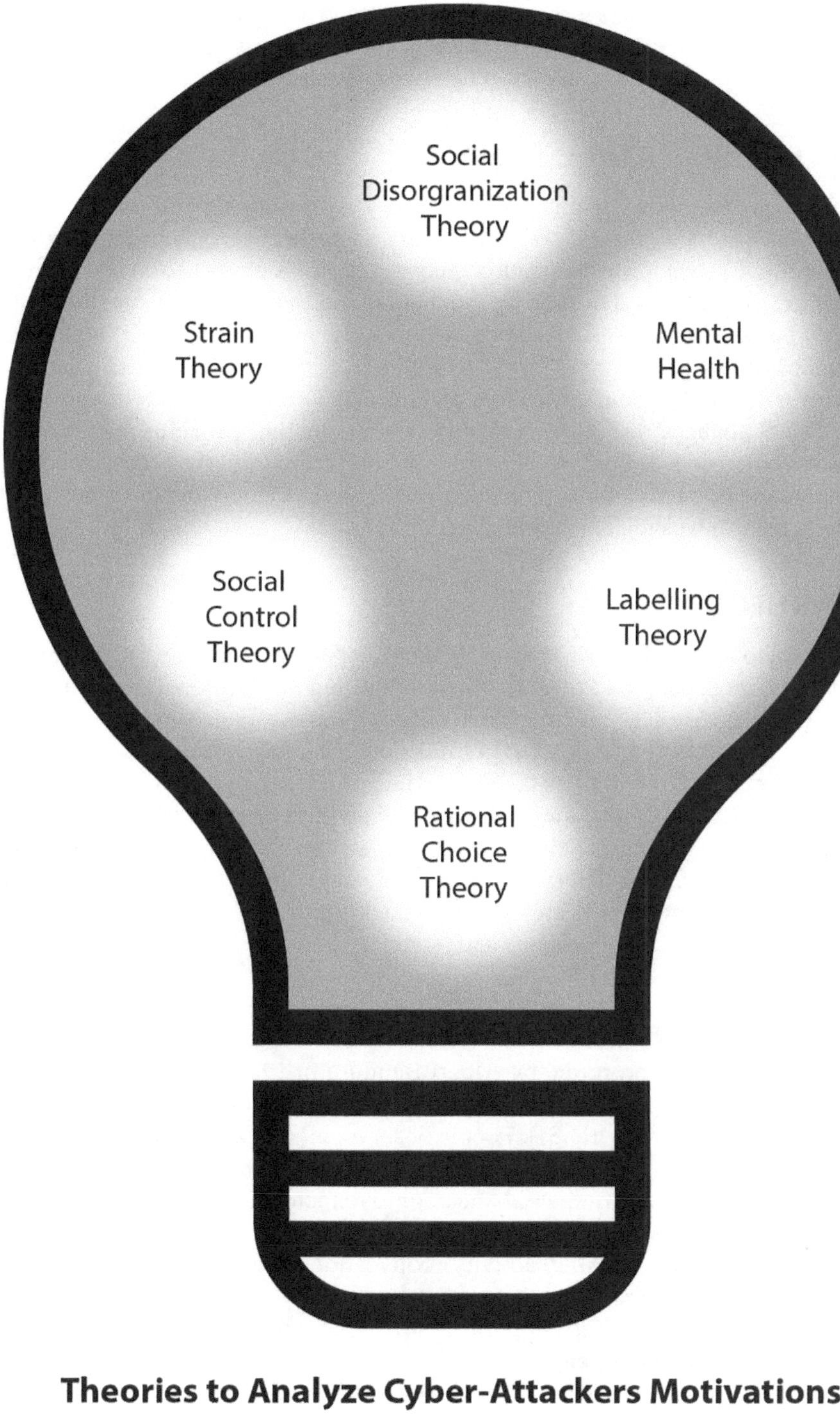

FIGURE 3.2 Principal theoretical frameworks to analyze cyberattacks.

The rational choice theory is used to understand insider threat attackers and explain the role of the target's attractiveness to the attacker as well as the effectiveness of defender protection strategies in deterring attacks [26,27].

SOCIAL DISORGANIZATION THEORY

In criminology, the social disorganization theory is used to explain some crimes. For example, why does the rate of crimes vary in different neighborhoods or cities? According to social disorganization theory, a person's physical and social environments are primarily responsible for behavioral choices [28]. Thus, a community with weak social structures, such as poor schools, vacant and vandalized buildings, high unemployment, and commercial and residential property, is more likely to have higher crime rates. As computer crimes and cybersecurity increase, structural and organizational changes, including laws and regulations that define crimes internationally, become needed. For instance, if the use of weapons of mass destruction or chemical weapons is internationally recognized as a crime, shouldn't the sabotage of water systems also be subject to regulation?

STRAIN THEORY

American sociologist Robert K. Merton 1938 developed the strain theory. According to this theory, most people have similar aspirations but have different opportunities or abilities. When people fail to achieve society's expectations through approved means such as hard work and delayed gratification, they may succeed through crime [29]. For example, a community's lack of income or education drives individuals to commit crimes. Consequently, the strain associated with low social control increases the likelihood of engaging in crime by reducing the social and emotional costs of criminal behaviors. Strain theory explains behavior such as cyberbullying among youth [30]. Other cybercrimes are computer hacking in many forms due to many motivational factors such as emotions and monetary gain.

SOCIAL LEARNING THEORY

Social learning theory, proposed by Albert Bandura in 1977, suggests that people develop the motivation to commit a crime and the skills to commit a crime through the people they associate with [31]. People observe, model, and imitate the behaviors, attitudes, and emotional reactions of others around them. Social learning theory considers how environmental and cognitive factors influence human learning and behavior. Consequently, people in an environment where crimes are part of their lives can expect criminal activity propagation. The example of people living in a community where gangs, terrorists, or criminal organizations are active is the primary motivation for people to learn that behavior. In addition, issues such as poverty, unemployment, poor housing, and political unrest lead to strain and status unrest in regions around the globe, motivating groups to commit cyber activities [32].

SOCIAL CONTROL THEORY

Travis Hirschi developed the social control theory in 1969, attempting to explain why individuals conform to societal norms and why others deviate from those norms. According to this theory, most people would commit a crime if not for society's control of individuals through institutions such as schools, workplaces, churches, and families. Thus, a strong social bond between individuals and society can prevent them from engaging in deviant behavior. Consequently, individuals who lack strong attachments to society, such as those who are socially isolated (e.g., gangs), are more likely to engage in deviant behavior [33,34].

Past studies suggest that computer security countermeasures such as security policies, systems, and awareness programs would effectively prevent computer abuse in organizations. This is because they are based on the general deterrence theory, which posits that when an organization implements countermeasures, this threatens abusers. However, computer abuse problems persist in many organizations despite these measures.

LABELING THEORY

Howard Becker labeling theory (1963) is an approach in sociology that individuals labeled as criminals modify the individual's self-concept and change how others respond to the labeled person [35]. People in power decide what acts are crimes, and the act of labeling someone a criminal is what makes him a criminal. Once a person is labeled a criminal, society takes away their opportunities, leading to more criminal behavior. That might be why, for example, people from some communities are more likely to be involved with policy interventions and more likely to lead to an arrest and severe punishment, accounting for the nature and seriousness of the offense [36].

Labeling theory explains how some whistleblowers might be subject to extreme punishment when authorities label them criminals. Some attackers might be labeled as criminals in one country, but the opposite is true in another.

MENTAL HEALTH

Sometimes, mental health issues may play a role in conducting cyberattacks. For example, individuals with antisocial personality or narcissistic personality disorder may be more prone to engaging in criminal behavior, including cyberattacks. Sometimes, individuals with mental health issues may use cyberattacks to gain attention or express their grievances.

CONCLUSION

In conclusion, understanding cyberattackers' motivations is fundamental to analyzing cyberattacks. Knowing cyberattackers' motivations helps organizations define their defense strategies and allocate resources accordingly to reduce the risk of loss from the occurrence of specific threats, better assess risk exposure, and better anticipate and defend against future attacks.

Overall, understanding the motivation of cyberattacks helps decision-making processes across various aspects of cybersecurity, from defense strategy development to incident response and regulatory compliance.

REFERENCES

1. Consumer Sentinel Network. (2021). "Data Book 2020," Consumer Sentinel Network/ Federal Trade Commission, February 2021. https://www.ftc.gov/system/files/documents/reports/consumer-sentinel-network-data-book-2020/csn_annual_data_book_2020.pdf.
2. N.C.U.A. (2019). "Ransomware Is a Serious and Growing Threat," NCUA, May 2019. https://www.ncua.gov/newsroom/ncua-report/2016/ransomware-serious-and-growing-threat.
3. Zetter K. (2014). "An Unprecedented Look at Stuxnet, the World's First Digital Weapon," Wired, November 3, 2014. https://www.wired.com/2014/11/countdown-to-zero-day-stuxnet/.
4. Doucet L. (2020). "Qasem Soleimani U.S.S. Kills Top Iranian General in Baghdad Air Strike," B.B.C., January 2020. https://www.bbc.com/news/world-middle-east-50979463.
5. Carole Cadwalladr C., Graham-Harrison E. (2018). "Revealed: 50 Million Facebook Profiles Harvested for Cambridge Analytica in Major Data Breach," The Guardian, March 17, 2018. https://www.theguardian.com/news/2018/mar/17/cambridge-analytica-facebook-influence-us-election.
6. Davies N. (2011). "Jonathan Rees: Private Investigator Who Ran Empire of Tabloid Corruption," The Guardian, March 2011. https://www.theguardian.com/media/2011/mar/11/jonathan-rees-private-investigator-tabloid.
7. Betschen A. (2018). "Shining a Light on Federal Law Enforcement's Use of Computer Hacking Tools," JustSecurity.Gov, September 19, 2018. https://www.justsecurity.org/60785/shining-light-federal-law-enforcements-computer-hacking-tools/
8. Sonnemaker T. (2020). "Law Enforcement Agencies Are Using a Legal Loophole to Buy up Personal Data Exposed by Hackers," Business Insider, July 8, 2020. https://www.businessinsider.com/police-buying-hacked-data-bypassing-legal-processes-2020-7.

8a. Louise Osborne L. (2015). "German State Buys USB Said to Hold Evidence of €600m in Tax Avoidance," The Guardian, November 7, 2015. https://www.theguardian.com/business/2015/nov/03/north-rhine-westphalia-tax-avoidance-usb-data-germany.

9. Ray M. (2022). "Edward Snowden: American Intelligence Contractor," Britanica, June 17, 2022. https://www.britannica.com/biography/Edward-Snowden.
10. MacAskill E., Thielman S., Thielman P. (2017). "WikiLeaks Publishes Biggest Ever Leak of Secret C.I.A.A. Documents," The Guardian, March 2017. https://www.theguardian.com/media/2017/mar/07/wikileaks-publishes-biggest-ever-leak-of-secret-cia-documents-hacking-surveillance.
11. Goodwin B. (2019). Computer Weekly: "Whistleblowers: James Glenn's Battle with Cisco Opens New Front on Cyber Security," Computer Weekly, August 2019. https://www.computerweekly.com/news/252468089/Whistleblowers-James-Glenns-battle-with-Cisco-opens-new-front-on-cyber-security.
12. Stouffer C. (2021). "Hacktivism: An Overview Plus High-Profile Groups and Examples," Norton, September 8, 2021. https://us.norton.com/internetsecurity-emerging-threats-hacktivism.html.
13. Swanson K., Kirsch II Th.L., Dunigan R. M. (2013). "Data Breaches in a Whistleblower's World: What You Should Know, Why You Should Know It," A.B.A.A. Criminal Justice Section Newsletter, Spring 2013. https://media.crai.com/wp-content/uploads/2020/09/16164509/Data-Breaches-in-a-Whistleblowers-World.pdf.
14. McClelland R. T. (2010). The Pleasures of Revenge. *The Journal of Mind and Behavior.* Vol. 31, No. 3/4, pp. 195–235.

15. McKee I. R., Feather N. T. T. (2008). Revenge, Retribution, and Values: Social Attitudes and Punitive Sentencing. *Social Justice Research*. Vol. 21, p. 138. https://doi.org/10.1007/s11211-008-0066-z.
16. Price M. (2009). Revenge and the People Who Seek It. *American Psychological Association*. Vol. 40, No. 6, p. 34. https://www.apa.org/monitor/2009/06/revenge.
17. Keeney M. et al. (2005). "Insider Threat Study: Computer System Sabotage in Critical Infrastructure Sectors," U.S.S. Secret Service and CERT Coordination Center/S.E.I. https://permanent.fdlp.gov/gpo151346/its_report_050516.pdf.
18. Taylor Ch. (2018). "World's Most Renowned Hacker on How Pranks Led to Prison," Irish Times, January 13, 2018. https://www.irishtimes.com/business/technology/world-s-most-renowned-hacker-on-how-pranks-led-to-prison-1.3353358#:~:text=Mitnick%2C%20who%20is%20from%20Los,huge%20money%2Dmaking%20venture.%E2%80%9D.
19. Bhattacharjee Sh. (2021). "'Did It for Fun,' Claims Suspected Hacker behind $600-Million Crypto Heist," CNBCTV18.Co, August 13, 2021. https://www.cnbctv18.com/cryptocurrency/did-it-for-fun-claims-suspected-hacker-behind-600-million-crypto-heist-10363611.htm.
20. Arthur Ch. (2013). "Burger King's Woes Continue after Twitter Account Gets Hacked," The Guardian, February 18, 2013. https://www.theguardian.com/technology/2013/feb/18/burger-king-twitter-account-hack.
21. Jayakumar Sh. (2021). Chapter 29: Cyber attacks by terrorists and other malevolent actors: Prevention and preparedness. In A. P. Schmid (Ed.), *Handbook of Terrorism and Preparedness*. The International Centre for Counter-Terrorism (I.C.C.T.), Washington, DC. https://icct.nl/app/uploads/2021/05/Handbook-ch-29-Jayakumar-Cyber-Attacks-by-Terrorists-and-other-Malevolent-Actors.pdf.
22. Zetter K. (2018). "Inside the Cunning, Unprecedented Hack of Ukraine's Power Grid," Wired, March 2016. https://www.wired.com/2016/03/inside-cunning-unprecedented-hack-ukraines-power-grid/.
23. Shane S., Perlroth N., Sanger D. E. (2017). "Security Breach and Spilled Secrets Have Shaken the N.S.A.A. to Its Core," New York Times, November 12, 2017. https://www.nytimes.com/2017/11/12/us/nsa-shadow-brokers.html.
24. Mandelcorn S., Modarres M., et al. (2013). "An Explanatory Model of Cyber-Attacks Drawn from Rational Choice Theory," Center for Risk and Reliability, University of Maryland, College Park, USA. https://drum.lib.umd.edu/handle/1903/14266.
25. Clarke R., Felson M. (2017). Activity and Rational Choice, Volume 5, chapter 14 pages, Routledge, New York.
26. Willison R., Siponen M. (2009). Overcoming the Insider: Reducing Employee Computer Crime through Situational Crime Prevention. *Communications of the ACM*. Vol. 52, No. 9, pp. 133–137.
27. Ransbotham S., Mitra S. (2009). Choice and Chance: A Conceptual Model of Paths to Information Security Compromise. *Information Systems Research, INFORMS*. Vol. 20, No. 1, pp. 121–139.
28. Bellair P. (2017). *Social Disorganization Theory*, Oxford Research Encyclopedia of Criminology. Retrieved 24 Dec. 2024, from https://oxfordre.com/criminology/view/10.1093/acrefore/9780190264079.001.0001/acrefore-9780190264079-e-253.
29. Agnew R. (1992). Foundation for a General Strain Theory of Crime and Delinquency. *Criminology*. Vol. 30, No. February 1992, pp. 47–88.
30. Patchin J. W., Hinduja S. (2011). Traditional and Non-traditional Bullying among Youth: A Test of General Strain Theory. *Youth & Society*. Vol. 43, No. 2, pp. 727–751.

31. Bandura A. (1977). *Social Learning Theory*, Englewood Cliffs, NJ: Prentice Hall; Bandura A. (1986). *Social Foundations of Thought and Action: A Social Cognitive Theory*, Englewood Cliffs, NJ: Prentice-Hall, Inc.; Bandura A., Ross D., Ross S. A. (1961). Transmission of Aggression through the Imitation of Aggressive Models. *Journal of Abnormal and Social Psychology*, Vol. 63, pp. 575–582.
32. Worden R. E., Shepard R. L., Mastrofski S. D. (1996). On the Meaning and Measurement of Suspects' Demeanor toward the Police: A Comment on "Demeanor and Arrest". *Journal of Research in Crime and Delinquency*. Vol. 33, No. 3, pp. 324–332.
33. Kempf K. L. (1993). The Empirical Status of Hirschi's Control Theory. *New Directions in Criminological Theory*. Vol. 4, pp. 143–185.
34. Hirschi T. (1969). *Causes of Delinquency*, Berkeley: University of California Press, Chapter 5, pp. 55–69. https://in.sagepub.com/sites/default/files/upm-binaries/36812_5.pdf.
35. Beker H. (1963). *Outsider: Studies in Sociology of Deviance*, New York: Free Press; Nickerson C., Becker's H. (2021). Labeling Theory. *Statistics*, Vol. 1, p. 2.
36. Shader M. (2003). "Risk Factors for Delinquency: An Overview," US Department of Justice, Office of Justice Program, NCJ 207540, January 2003, Page 11. https://www.ojp.gov/ncjrs/virtual-library/abstracts/risk-factors-delinquency-overview.

Part Two

Cyberwar and Cyberespionage

- SolarWinds Case
- Pegasus Spyware
- Google Chrome Extension Spyware
- Marriott International

4 SolarWinds

ABSTRACT

This chapter examines the SolarWinds Case, what we know about it, the data compromised, and the case from a human factors point of view.

INTRODUCTION

SolarWinds is a U.S. information technology management software company for network management that provides IT services to all five branches of the U.S. government, including the military, the State Department, the National Security Agency, and the Office of the President of the United States. SolarWinds also serves U.S. Fortune 500 companies, such as the top 10 U.S. telecommunications providers.

In December 2020, it was reported that a nation-state, suspected to be Russia, subverted SolarWinds' monitoring software updates through a sophisticated, targeted, and manual supply chain attack. Subsequently, the investigation revealed further details about the extent of the damage, which was unprecedented, making it the largest attack of its kind ever recorded [1].

After a brief description of the case, this chapter looks at what we know about this case, the data compromised, and analyzes the case from human factors point of view.

CASE DESCRIPTION

Orion is a monitoring and management Software as a Service (SaaS) tool of SolarWinds. Orion is used by more than 300,000 clients, including Information technology (IT) companies and governments. Since Orion is a network monitoring system, it has privileged access to IT systems to obtain log and system performance data [2]. Cyberattackers inserted malicious code into the Orion system and accessed all organizations that use Orion to monitor and manage their systems.

The SolarWinds attack was used to break into another computer network. In the technological world, this type of attack is called a *supply chain attack* or *third-party attack*. Supply chain attacks target vendors and suppliers instead of directly targeting a specific business. Consequently, it is harder to detect and prevent if the vendors don't maintain strict cybersecurity policies and use the best tools. This attack aimed to access source codes, build processes, or update mechanisms by infecting legitimate apps for the purpose of distributing malware.

In this case, attackers used SolarWinds software to infiltrate other companies by exploiting a vulnerability in SolarWinds Orion. This type of attack is similar to using one Google Home to unlock somebody's home door like a thief.

FireEye, another cybersecurity company, first discovered the SolarWinds case then came into the limelight in early December 2020 [3].

DOI: 10.1201/9781003461739-6

According to the reports, FireEye was investigating its stolen Red Team toolkit and discovered that the SolarWinds Orion software update distribution mechanism was used to spread malicious code to organizations using the software [4].

Soon after, the extent of the damage was revealed to be very large and exponentially growing. First, The Wall Street Journal reported that the U.S. Commerce and Treasury Departments, the Department of Homeland Security (DHS), the National Institutes of Health (NIH), and the State Department were all affected. Then, gradually, the list of victims' organizations became larger and larger, followed by the involvement of the U.S. Federal Bureau of Investigations (FBI), the Cybersecurity and Infrastructure Security Agency (CISA), the Office of the National Director of Intelligence (ODNI), and the National Security Agency (NSA) [5].

Later, it was discovered that the cyberattackers also gained access to the email accounts of Donald Trump's Department of Homeland Security and various other employees in his office.

It appeared then that the attack was more widespread and began much earlier than expected, estimated to be sometime in March 2020, which meant the attack had been underway for months before its detection in December 2020.

Since multiple networks had been penetrated, secure systems were expensive and challenging. President Trump's homeland security officer Tom Bossert said that it could be years before the networks are protected again. With access to government networks, hackers could "destroy or alter data and impersonate legitimate people" [6].

WHAT WE KNOW ABOUT THIS CASE

As per the attack description, it seems the attackers carefully planned for a long time before executing. They were patient and took their time to take small steps and not raise suspicion. Based on the attack timeline, they started with non-government organizations and private IT companies to grab their hold inside the system [7].

After gaining access to Orion's internal system, they digitally signed the certificates for the new software updates, which were supposed to be installed by all the customers, including all the major organizations. Instead, the update gave cyberattackers access to users' systems when they installed the new update [3]. Then, the attacker's code created a backdoor to customers' information technology systems, which cyberattackers then used to install even more malware that helped them spy on companies and organizations.

INFORMATION COMPROMISED

The primary motive of the attackers remains unclear. However, from the analysis, we can identify that the cyberattackers were looking to get confidential government data, including nuclear documents, politicians' emails, and other trade secrets from huge IT companies.

According to U.S. President Donald Trump, Chinese cyberattackers were behind this attack. Later, federal investigators and cybersecurity experts found that the Russian Intelligence Services (SVR) was behind the attack [3,5].

Some of the victims of this supply-chain attack are as follows:

- Energy Department (DOE) and National Nuclear Security Administration (NNSA).
- FireEye: Testing and hacking toolkits were stolen.
- Microsoft: Important source code was stolen.

WHAT CYBERATTACKERS CAN DO WITH HACKED DATA

As rumors continue circulating regarding whether Russian Intelligence was behind the attack [3,5], if the cyberattackers were not from the government, they could sell personal data, such as passwords and trade secrets, to other countries or on the black market. The compromised nuclear files can be used against all U.S. adversaries. The compromised data hacked could result in a significant loss in nuclear expenditure by the U.S. government. The source codes of companies like Microsoft can crash the company's stock prices. Exposing the confidential technology of private companies could create a recession in the U.S. economy. Additionally, the exposure of politicians' emails could overturn the ruling party.

Given the difficulty of determining the extent of information and software codes obtained, modified, or altered by the attackers from the start of the attacks until they were discovered, victim organizations must update their security systems with new methods to preempt these attacks and consider that their information may be in the hands of adversaries, especially in the case of government agencies [8,9]. The information includes country infrastructure systems and confidential information. As a report from the New York State Department of Financial Services indicates,

> The SolarWinds Attack should serve as a wake-up call. Russian hackers opened back doors into thousands of organizations through a single vector, including almost 100 companies in New York's financial services industry. Although none of the networks of the DFS-regulated companies were actively exploited, the SolarWinds Attack highlights the financial services industry's vulnerability to supply chain attacks. The SolarWinds attack confirms that cyber risks are a threat not just to consumers and individual companies but also to the stability and soundness of our entire financial services industry. This is an existential threat, and we urge the industry to treat it as such. [10]

HUMAN FACTORS ANALYSIS

The SolarWinds attack, with all its technical and complex timelines, has two components.

The first technical aspect includes malicious code added into the Orion system and access to all organizations that use Orion to monitor and manage the systems. That component is different from the focus of this chapter and book. Interested readers might refer to the SolarWinds investigation report [11] and CrowdStrike team report [12,13] for the technical aspect of the malware and the attack.

The second human factors component includes the human behavior that allowed these types of attacks to happen, as well as the needed behavior modification of IT professionals after this attack. Based on what we know, several areas in which human agent behavior was crucial in this attack were detection, penetration, and monitoring after the attack. Therefore, this cyberattack case has several areas in which human behavior needs to be considered.

DETECTION

As mentioned above, FireEye discovered the SolarWinds case in early December 2020 [4]. Discovery occurred when FireEye discovered the hackers through a device registered with an employee's credentials. FireEye's security system alerted the employee and the company's security team, informing them that a new device had just been registered to its Multi-Factor Authentication (MFA) system as if it belonged to the employee. The alert prompted FireEye to investigate. Since a registered employee device to FireEye's MFA system could receive the employee's unique access codes, it triggered FireEye to discover how the hackers obtained the employee's credentials to register their devices. Consequently, this investigation uncovered the SolarWinds breach into their network.

The usage of MFA was effective in starting an investigation. In this case, the human agent monitoring system worked well to investigate the alert that led to the discovery of the attacks. It is essential to acknowledge that it is unclear how attackers may have obtained the employee's credentials. No further information about any evidence of FireEye employees' credentials to the attackers is reported. The attackers might have obtained the employee credential once they got into the FireEye system through SolarWinds. No matter how the attackers got the employee's credentials, it is a fact that the starting point for an attack is acquiring credentials, making this a human issue.

ATTACK

Trust

IT professionals are encouraged to update software promptly, mainly when there is a security update. The security update is built and released by trusted vendors, signed, and certified. IT professionals trusting the security update most probably did not entrust the security update from a security vendor. Thus, they updated their apps, unaware that the update is infected with malicious code. The malicious code was also run with trust and permissions.

Contrary to this suggested habit, in this case, the customers who didn't update the software were not harmed. However, this behavior could influence IT professionals' behavior in future situations in the cyberspace domain. Prompt updating behavior might result in waiting to update the system, which might again have another security vulnerability since many people are inclined not to update software, as customers who did not update the software were not harmed. This attitude could worsen the current situation in the cyberspace domain [5].

Outlier Effect

Attackers in the SolarWinds case act in minimal steps to not raise any suspicion. People tend to disregard an outlier that is a data point, or system monitoring differs significantly from other observations. For example, it might be allocated to variability in the measurement, or it may indicate system error. In this case, security operators who might have been excluded from an investigation did not notice the attack, so their behavior might be due to misjudgment.

BEHAVIOR AFTER ATTACK

Companies have started making massive reforms in cybersecurity procedures. For example, they have included a new approach of "Assume Breach Mentality" that companies assume by managing and organizing their confidential data in a way that cyberattackers are already in their system [5].

The number of potential victims is significant, given the popularity of some apps. For example, a case occurred where a free file compression app was poisoned and deployed to customers in a country where it was the top utility app.

With open-source development and increasing third-party microservices, more focus should be placed on improving app security to avoid such supply chain attacks, instead of halting their development.

CONCLUSION

The consequences of this attack are enormous. It should be assumed that cyberattackers got access to all the critical data from the U.S. governments and companies that were exposed. However, this assumption required a difficult task for all the affected organizations [6,14].

The other main issue is trusting other services requiring every access, even from the service provider, which needs to be evaluated with care and nothing be considered fully charged. This case illustrates the essential need of human monitoring and investigation in automating systems for IT professionals [15,16].

REFERENCES

1. Temple-Raston D. (2021). "A 'Worst Nightmare' Cyberattack. The Untold Story of the SolarWinds Hack," NPR, April 18, 2021. https://www.npr.org/2021/04/16/985439655/a-worst-nightmare-cyberattack-the-untold-story-of-the-solarwinds-hack.
2. Korolov M. (2021). "What Is a Supply Chain Attack? Why to be Wary of Third-Party Providers," CSO Online, February 4, 2021. https://www.csoonline.com/article/561323/supply-chain-attacks-show-why-you-should-be-wary-of-third-party-providers.html.
3. Mia Jankowicz M., Davis Ch. (2020). "These Big Firms and U.S. Agencies All Use Software from the Company Breached in a Massive Hack Being Blamed on Russia," Business Insider, December 14, 2020. https://www.businessinsider.com/list-of-companies-agencies-at-risk-after-solarwinds-hack-2020-12.
4. FireEye. (2020). "Highly Evasive Attacker Leverages SolarWinds Supply Chain to Compromise Multiple Global Victims with SUNBURST Backdoor," FireEye, December 13, 2020. https://www.fireeye.com/blog/threat-research/2020/12/evasive-attacker-leverages-solarwinds-supply-chain-compromises-with-sunburst-backdoor.html.
5. Volz D., McMillan R. (2020). "Suspected Russian Hack Said to Have Gone Undetected for Months," Wall Street Journal, December 15, 2020. https://www.wsj.com/articles/suspected-russian-hack-said-to-have-gone-undetected-for-months-11607974376.

6. Baker P. (2021). "The SolarWinds Hack Timeline: Who Knew What, and When?" CSO Online, April 5, 2021. https://www.csoonline.com/article/3613571/the-solarwinds-hack-timeline-who-knew-what-and-when.html.
7. Bossert Th. P. (2020). "I Was the Homeland Security Adviser to Trump. We're being Hacked," New York Times, December 16, 2020. https://www.nytimes.com/2020/12/16/opinion/fireeye-solarwinds-russia-hack.html?action=click&module=Opinion&pgtype=Homepage.
8. Jibilian I. (2021). "The U.S. Is Readying Sanctions against Russia over the SolarWinds Cyber-Attack. Here's a Simple Explanation of How the Massive Hack Happened and Why It's Such a Big Deal," Business Insider, April 15, 2021. https://www.businessinsider.com/solarwinds-hack-explained-government-agencies-cyber-security-2020-12.
9. Evstatieva M. (2021). "A 'Worst Nightmare' Cyberattack: The Untold Story of the SolarWinds Hack," NPR, April 16, 2021. https://www.npr.org/2021/04/16/985439655/a-worst-nightmare-cyberattack-the-untold-story-of-the-solarwinds-hack.
10. New York State Department. (2021). "Report on the SolarWinds Cyber Espionage Attack and Institutions' Response," New York State Department, April 2021, Page 13. https://www.dfs.ny.gov/system/files/documents/2021/04/solarwinds_report_2021.pdf.
11. CIS. (2021). "The SolarWinds Cyber-Attack: What You Need to Know," Center for Internet Security, March 15, 2021. https://www.cisecurity.org/solarwinds/.
12. Ramakrishna S. (2021). "New Findings from Our Investigation of SUNBURST," SolarWinds, January 2021. https://orangematter.solarwinds.com/2021/01/11/new-findings-from-our-investigation-of-sunburst/.
13. CrowdStrike Intelligence Team. (2021). "SUNSPOT: An Implant in the Build Process," Crowdsrike Blog, January 2021. https://www.crowdstrike.com/blog/sunspot-malware-technical-analysis/.
14. FireEye. (2022). "Highly Evasive Attacker Leverages SolarWinds Supply Chain to Compromise Multiple Global Victims with SUNBURST Backdoor," FireEye, December 13, 2020, updated May 2022.
15. Bing Ch. (2020). "Suspected Russian hackers Spied on U.S. Treasury Emails – Sources," Reuter, Business Report, December 13, 2020. https://www.reuters.com/article/us-usa-cyber-treasury-exclsuive/suspected-russian-hackers-spied-on-u-s-treasury-emails-sources-idUKKBN28N0PG?edition-redirect=uk.
16. Higgins K.J. (2021). "FireEye's Mandia: 'Severity-Zero Alert' Led to Discovery of SolarWinds Attack," Dark Reading, January 7, 2021. https://www.darkreading.com/threat-intelligence/fireeye-s-mandia-severity-zero-alert-led-to-discovery-of-solarwinds-attack.

5 Pegasus Spyware

Abstract

This chapter examines the Pegasus Spyware that infected billions of phones running either iOS or Android operating systems, what we know about it, the compromised data, and the case from a human factors point of view.

INTRODUCTION

The mobile phone is a device that almost everyone worldwide carries in their pocket. The ubiquity of mobile phones is undeniable, as they have become a crucial component of modern society, relied upon by individuals ranging from high-profile figures, such as heads of state and company executives, to everyday people. They play an indispensable role in both our personal and professional lives. Consequently, the smartphone has become a compelling device to be targeted by cyberhackers with different motivations, from financial to spying agencies. Attackers' primary way of dealing with mobile phones is through Spyware applications. Installing spyware on a mobile phone enables attackers to access a victim's Short Message Service (SMS) messages and contact details, record their call logs, or remotely activate a device's microphone and camera to secretly capture audio, video, and image content.

Pegasus is the hacking software developed, marketed, and licensed to governments worldwide by the Israeli company NSO Group. NSO's mass surveillance software can infect phones and other devices without using that device and the person ever knowing. NSO claims that this software is intended for government agencies to use in the prevention and investigation of crime and terrorism [1].

However, in August 2016, a human rights activist's failed application installation attempt on the iPhone led to an investigation revealing details about the spyware, its abilities, and the security vulnerabilities it exploited. According to several investigations, Pegasus spyware has been used to surveil anti-regime activists, journalists, and political leaders from several nations worldwide [2,3]. Also, various governments use the software for nefarious purposes, including multiple organizations finding Pegasus on the phones of people who are not terrorists, but instead social activists and others who criticize their governments [4,5].

Pegasus can infect billions of phones running either iOS or Android operating systems. It is a sophisticated application that can transform a smartphone into a 24-hour surveillance spying machine using the phone camera and microphone to record conversation sounds and video, including locations [6].

DOI: 10.1201/9781003461739-7

CASE DESCRIPTION

Pegasus's earliest version of spyware, developed by NSO Group and captured by researchers in 2016, infected phones through spear-phishing—text or emails that trick a target into clicking on a malicious link. An investigation by the Guardian and 16 other media organizations revealed that human rights activists, journalists, and lawyers worldwide had been targeted by authoritarian governments using hacking software sold by the Israeli surveillance company NSO Group [7]. The numbers of more than 180 journalists were monitored, including the data of reporters, editors, and executives at the *Financial Times*, *CNN*, the *New York Times*, *France 24*, *the Economist*, *Associated Press*, and *Reuters*. Later, Pegasus spyware was found on the phone of a jailed critic of Narendra Modi, the Indian Prime minister [8–10]. In January 2022, The Indian Express revealed that India bought Pegasus as part of a defense deal with Israel in 2017 [11]. Pegasus spyware was also found on Jamal Khashoggi, a Saudi dissident whom agents of the Saudi government assassinated [12,13]. Through a list of 50,000 leaked phone numbers, *the Guardian* also discovered that potential governmental targets were connected using NSO Pegasus software.

WHAT WE KNOW ABOUT THIS CASE

The earlier version of Pegasus was installed on users' smartphones by spear-phishing, meaning that the users needed to click a link that triggered the software installation or through a wireless network near a target. However, the more recent version of Pegasus can be installed on smartphones with a missed call on WhatsApp or simply through a message to a user's phone. In both cases, no notification is produced, and a user would not notice that the spyware has been installed on users' phones [14].

Pegasus spyware can be inserted into any mobile device to avoid encryption altogether. When installed on a target device, the software can turn on microphones, track location, and attack data without the person's knowledge. Essentially, Pegasus attacks zero-day holes in a product, which are loopholes and vulnerabilities within the hardware or software of said device that have not yet been seen by the companies. The company has many clients, but its primary clients are intelligence agencies, law enforcement, and the government [6].

After the Pegasus spyware is installed, it can harvest any data from the device and transmit it back to the attacker. Consequently, attackers can access users' photos, call logs, social media posts, etc. This spyware can also activate cameras and microphones for real-time surveillance without the user's permission or knowledge.

INFORMATION COMPROMISED

According to technical analyses [15], this spyware takes extensive amounts of personal data for surveillance purposes, including access to your camera, microphone, and location, without someone knowing [16].

It turns the device's camera and microphone on without the owner's knowledge and can listen to conversations in the room or area where the device is located. The spyware also tracks calendars, photos, messages, and other contacts.

The spyware can only access the data stored on the device. However, attackers may be able to gain access to additional data by exploiting shared data between the victim and other individuals.

WHAT CYBERATTACKERS CAN DO WITH HACKED DATA

The victims of Pegasus spyware are essentially tracked. They will be under total surveillance for what they say, read, see, or send. Their location is tracked and known, and the victims' spying might continue indefinitely, depending on the attackers' needs or until victims finally find out they've been surveilled.

If the attacker aims to cause physical harm to the person or their surroundings, they have their location and make them and the people they meet valuable assets. From a tracking point of view, the attackers can use this information to blackmail, plan other cyberattacks, and receive sensitive information to intersect or move against the person focused on the attack.

HUMAN FACTORS ANALYSIS

From a human factors point of view, victims cannot do very much to combat such an attack since the spyware is installed through a zero-click attack and without clicking a phishing link. However, several human factors and human–computer interaction principles are used in the design of this spyware, which is essential for analyzing this case.

Ease of Use

Spyware designers have plentiful resources that they use to design the spyware with usability and ease of use in mind. Once the application is purchased to attack a victim, only a contact address is needed; the victim will not need to do anything, and the phone will be under the attacker's control. Users only sometimes analyze missed calls or text messages when they receive many spam calls or texts. Also, people rarely look at the applications installed on their mobile phones to see if an unfamiliar application has been downloaded, and it's also unusual that they discover suspicious behavior on their device. Thus, there should be greater efforts put into warning and teaching mobile phone users to check their phones for such issues frequently.

Privacy

Perceived security generally refers to consumers' beliefs that their personal information, such as personal data, will not be viewed, stored, or used by unauthorized third parties [17]. Users' perceived security is essential when using communication applications and many people use WhatsApp because they know it has end-to-end data encryption.

Thus, user trust in the safety of an application is relatively high. Although not very public news, this could cause a lack of confidence in WhatsApp users, as the user interface and system need to warn the user when a threat is detected. Users' privacy and trust continue to be an important vulnerable point for attacks to occur.

Trust

Unless a forensic analysis of a phone is done, users will never know if their device has been infected with Pegasus. Consequently, device manufacturers should include the ability to scan for tracking applications to enhance user confidence and build and integrate methods that help the device stay as protected as possible [17,18].

Awareness

People using any digital communication should always consider or keep in mind that they might be subject to this attack. Interestingly, many movements and security agencies are customers of such technologies. Thus, without appropriate laws and regulations, citizens remain vulnerable.

CONCLUSION

Despite media reports, lawsuits by Apple Computer, and involvement from various governments, it is unlikely that security agencies in many known or unknown countries will cease using technologies such as Pegasus Spyware. In fact, these technologies may continue to advance unless adequate laws and regulations are put in place [19].

REFERENCES

1. Federam J. (2022). "NSO Turns to U.S. Supreme Court for Immunity in WhatsApp Suit," AP NEWS, April 11, 2022. https://apnews.com/article/us-supreme-court-technology-business-spyware-lawsuits-3a2cdcfdac224647bd65e95fd57395d5 (accessed April11, 2022).
2. Levinson Ch. (2020). "With Israel's Encouragement, NSO Sold Spyware to UAE and Other Gulf States," Haaretz, August 25, 2020. https://www.haaretz.com/middle-east-news/.premium-with-israel-s-encouragement-nso-sold-spyware-to-uae-and-other-gulf-states-1.9093465.
3. Amnesty International. (2021). "Forensic Methodology Report: How to Catch NSO Group's Pegasus," Amnesty, July 18, 2021. https://www.amnesty.org/en/latest/research/2021/07/forensic-methodology-report-how-to-catch-nso-groups-pegasus/
4. Bergman R., Mazzetti M. (2022). "The Battle for the World's Most Powerful Cyberweapon," The New York Times, January 28, 2022. https://www.nytimes.com/2022/01/28/magazine/nso-group-israel-spyware.html (accessed April 11, 2022).
5. Perlroth N. (2016). "How Spy Tech Firms Let Governments See Everything on a Smartphone," The New York Times, September 2, 2016. https://www.nytimes.com/2016/09/03/technology/nso-group-how-spy-tech-firms-let-governments-see-everything-on-a-smartphone.html.
6. Pegg D., Cutler S. (2021). "What Is Pegasus Spyware, and How Does It Hack Phones?" The Guardian, July 18, 2021. https://www.theguardian.com/news/2021/jul/18/what-is-pegasus-spyware-and-how-does-it-hack-phones.
7. Kirchgaessner S. et al. (2021). "Revealed: Leak Uncovers Global Abuse of Cyber-Surveillance Weapon," The Guardian, July 18, 2021. https://www.theguardian.com/world/2021/jul/18/revealed-leak-uncovers-global-abuse-of-cyber-surveillance-weapon-nso-group-pegasus.
8. Shankland S. (2021). "Pegasus Spyware on State Department Phones: What You Need to Know," CNET, December 3, 2021. https://www.cnet.com/tech/mobile/pegasus-spyware-on-state-department-phones-what-you-need-to-know/.

9. Bing Ch., Joseph M. J. (2021). "U.S. State Department Phones Hacked with Israeli Company Spyware – Sources," Reuter, December 3, 2021. https://www.reuters.com/technology/exclusive-us-state-department-phones-hacked-with-israeli-company-spyware-sources-2021-12-03/.
10. Kirchgaessner S., Michael S. M. (2022). "Pegasus Spyware Found on Phone of Jailed Critic of Narendra Modi," December 17, 2021. https://www.theguardian.com/news/2021/dec/17/pegasus-spyware-found-on-phone-of-jailed-critic-of-narendra-modi-.
11. Kaushik K. (2022). "India Bought Pegasus as Part of Defense Deal with Israel in 2017: NYT," The Indian Express, January 30, 2022. https://indianexpress.com/article/india/india-bought-pegasus-defence-deal-israel-2017-nyt-7746655/.
12. PBS. (2018). "How NSO Group's Pegasus Spyware Was Found on Jamal Khashoggi's Fiancé's Phone," FRONTLINE, www.youtube.com, July 21, 2018. https://www.youtube.com/watch?v=a2BIYWHdfTE (accessed April 11, 2022).
13. BBC. (2021). "Pegasus: Who Are the Alleged Victims of Spyware Targeting?" BBC, July 22, 2021. https://www.bbc.com/news/world-57891506.
14. Bhankiran G. B. (2021). "What Is Pegasus? How Surveillance Spyware Invades Phones," Scientific American, August 9, 2021. https://www.scientificamerican.com/article/what-is-pegasus-how-surveillance-spyware-invades-phones/#:~:text=Once%20installed%2C%20Pegasus%20can%20theoretically,logs%20and%20social%20media%20posts.
15. Max Bazaliy M., Flossman M. et al. (2016). "Technical Analysis of Pegasus Spyware: An Investigation into Highly Sophisticated Espionage Software," Lookout. https://info.lookout.com/rs/051-ESQ-475/images/lookout-pegasus-technical-analysis.pdf.
16. OCCRP. (2021). "How Does Pegasus Work?" OCCRP, July 18, 2021. https://www.occrp.org/en/the-pegasus-project/how-does-pegasus-work (accessed April 11, 2022).
17. Flavián C., Guinalíu M. (2006). Consumer Trust, Perceived Security and Privacy Policy: Three Basic Elements of Loyalty to a Web Site. *Industrial Management & Data Systems.* Vol. 106, No. 5, pp. 601–620. https://doi.org/10.1108/02635570610666403.
18. Zarouali B. et al. (2021). WhatsApp Marketing: A Study on WhatsApp Brand Communication and the Role of Trust in Self-Disclosure. *International Journal of Communication, [S.l.],* Vol. 15, p. 25. https://ijoc.org/index.php/ijoc/article/view/15365/3318 (accessed August 11, 2022).
19. Apple. (2021). "Apple Sues NSO Group to Curb the Abuse of State-Sponsored Spyware," Press Release, November 23, 2021. https://www.apple.com/newsroom/2021/11/apple-sues-nso-group-to-curb-the-abuse-of-state-sponsored-spyware/.

6 Google Chrome Extension Spyware

Abstract

This chapter, after a brief description of the Google Chrome Extension Spyware case, looks at what we know about this case, the data compromised, and analyzes the case from a human factors point of view.

INTRODUCTION

Extensions in web browsers like Google Chrome and Mozilla Firefox present a novel opportunity for hackers to compromise the users' data. Browser extensions pose a significant threat to user data because even the most widely used on-host installed anti-virus security software struggles to detect malicious activity within web browsers. This issue is especially critical because a web browser is the most used application by nearly all computer users. Therefore, if a hacker gets access to the web browser on the system, it is like having keys to all users' systems. Everything, including the user's email, screenshots, clipboard, credential tokens stored in cookies or parameters, and user keystrokes, is susceptible to theft or modification.

Google's lack of oversight of its Chrome browser resulted in several extensions going undetected when they surreptitiously compromised the data of millions of users. These extensions performed a legitimate function to gain the users' confidence but then siphoned off data from their computers, including email addresses, payroll stubs, and credit card information.

According to a report, on June 17, 2020, sophisticated spyware evaded anti-virus software by disguising its nefarious operations whenever the anti-virus software was running. It is reported to be the most significant security vulnerability discovered in the Google ecosystem of products, as over 100 spyware extensions were downloaded 32 million times, affecting millions of people. This case is an important cause of concern because the spyware extensions were downloaded from the official Google Chrome Extension Library and not unofficial sources. This incident forced Google to amplify its security scanning of browser extensions and use a combination of human interaction and machine learning to detect such vulnerabilities more robustly [1].

After a brief description of the case, this chapter looks at what we know about this case, the data compromised, and analyzes the case from a human factors point of view.

DOI: 10.1201/9781003461739-8

CASE DESCRIPTION

While browser extensions are generally considered harmless, attackers have found ways to exploit the approval process of extension libraries, even with non-malicious applications, to gain access to user data. These approved applications can push malicious updates to the software. One of the ways to identify this potential vulnerability is to look at the permissions requested by the application carefully. For example, one of the malicious extensions was harmlessly named "Management Extension—Internal Chromium Extension" and requested pervasive permissions. It asks for enough capabilities to gather what users type in the browser, scrape tokens from public and internal corporate sites, content from the user's clipboard, save screenshots, and extract other susceptible sources of information.

A browser extension is a small software module to customize a web browser. All browsers allow extensions that include user interface enhancement, cookie management, advertisement blocking to remove unwanted website elements such as pop-up ads, and other feature enhancements such as auto-play for online videos. However, browser extensions are different from plugins and have different module types. Plugin software adds new functions to a host program without altering it.

Some browser extensions can be installed into the victim's machine. Attackers design professional-looking web pages to offer attractive features, encouraging users to download extensions while performing other hidden activities through sophisticated approaches. They also make previously installed malicious extensions install new malicious plugins. Security researchers have noticed and documented a few patterns to identify such dubious extensions in this case. The following signs can be used to distinguish fraudulent extensions:

- An unknown organization develops the extension or has an unreliable reputation in the industry.
- They are almost always very highly rated in the reviews section. The applications often have a five-star rating but generic and repetitive customer testimonials. A high rating lets the hackers list their malicious extensions in a favorable placement on the Extension Library.
- Slightly modified versions of these extensions are added multiple times to the Extension Library to evade detection.
- Multiple extensions have the same logo but with different color schemes.

Awake Security conducted security research and found 111 Chrome extensions to be malicious. Of these 111 extensions, 79 were available for download from the official Google Chrome Extension Library. A conservative estimate states that these extensions have been downloaded 32,962,951 times [2,3].

WHAT WE KNOW ABOUT THIS CASE

Awake Security, the firm that uncovered this security incident, provided the details about how these attacks happened and the groups that seem to be behind them. According to Awake Security, the common factor is that the offending domain name was Gal Communication Limited. It was responsible for creating and verifying these subdomains, as well as providing Domain Name System (DNS) hosting services to these domains. Awake Security analyzed 26,079 domains created by Gal Communication Limited as of early 2020 for this report. They started profiling the traffic profiles of these domains and looked for ways in which the users were redirected to exploitative landing pages. Their study concluded that 15,160 domains were either directly compromised or hijacked by Gal Communication Limited after expiration. The artifacts analyzed helped uncover this attack:

1. Malware source code, Command-and-control servers, and malicious landing pages
2. JavaScript code and HTML elements of the user-facing landing pages
3. Server versions and types
4. Response codes along with sent data
5. Error messages and their structure
6. Where the domain is hosted and the name server characteristics
7. Redirection chain properties
8. Tracking, web analytics elements, and cookies used in landing pages

This group of hackers was believed to be working on a larger conspiracy on a nation-state scale. The Arista NDR Threat Research Team suspected this because the domain names used for the Chrome extension attack are the same domain names associated with other large-scale attacks. Awake Security also found a habitual way of operating, representing a discernible pattern with these nation-states' sponsored spyware missions [3]:

1. A Vietnamese government-backed rogue hacker network known as Ocean Lotus (previously responsible for injecting malicious JavaScript code into users' browsing sessions and sniffing sent data) was noticed distributing applications and extensions in Google Chromes Extension Library and other well-known App distribution platforms like the Play Store and Mozilla Extension Store.
2. Elaborate phishing schemes targeting government and army officials from Pakistan used the same domain names as the Chrome extension attack.
3. The espionage-focused hacking group Confucius distributed malicious real-time chat applications that begin tracking the sent messages once a particular keyword of interest is noticed [4].

INFORMATION COMPROMISED

Since the hackers weaponized the browser extensions, they had a large attack radius. The web browser runs with almost administrator-level privileges with unrestricted access to the file system, networking, and memory resources on any operating system. Due to false advertising, the users believed these extensions would help them in their daily utility tasks like converting files, monitoring resources, document viewing, etc. As these extensions performed varied functionalities, the users gave them extensive permissions to access all the resources on the device.

Extensive permissions allowed the extensions to secretly transfer the user's browsing history and locally saved login credentials for accessing public websites and internal corporate portals. These extensions also have access to sessions in all of the browsers open tabs, which means the information sent and received by the user during their entire browsing session is available to the extensions. Some extensions also installed keyloggers on the user's device, which leaked all the user's passwords and private keys.

WHAT CYBERATTACKERS CAN DO WITH HACKED DATA

In the Google Chrome Extension, spyware attackers stole considerable user data. Consequently, they can use private data and files to gather confidential information about a victim's employer. The data such as trade secrets or intellectual property information is consequential and can make or break an organization. The victims are also personally exposed to the attacker, wherein they can be extorted or blackmailed by the attacker in possession of their private and compromising media. The credentials recovered from the key loggers can be used to take over the victim's other social media accounts or even banking portal access. Enough information is available to hijack the identity of the victim. These identity data sets are valued at hundreds of thousands of dollars on the dark web. The stolen information may also be used to conduct elaborate phishing scams on the victims. As the attackers have the victim's private information, they can craft a personalized phishing email that can easily trick the victim.

HUMAN FACTORS ANALYSIS

Google Chrome Extension Spyware used different human behavior patterns for this attack. The attackers used a combination of different behavior patterns that included the attractiveness of a free application, people's behavior of not reading terms of permission, access, and use, and trusting all types of extensions.

Power of Free Application

People are very much attracted to "free" items or "sales" in their behavior in purchasing. "Buy one get a second for free," "free sample," and "50% discount" are always attractive to people. Therefore, offering "free apps" is very popular among many people. Many legitimate apps or browser extensions offer to use an app for free in exchange for using the apps. Then, when users start using the app, they bombard them with ads. This behavior is widespread; generally, people do not necessarily check the developer's background or the company offering those apps. More importantly, they do not read the terms of usage and app permission list [5].

Research shows that the attraction of people to a free item is often irrational, particularly in this case for free games. However, as Dan Ariely calls it, they still do it because of the "Power of Free" [6]. In cybersecurity, attackers use this behavior to motivate them to download the apps.

Permission Access & Term of Use

The Google Chrome Extension spyware case leads us to believe humans must better identify such suspect software. Furthermore, since wanting free or discounted things is a widespread behavior, it would be important that web browsers, operating systems, and app stores provide warnings and remind people what to check for before downloading any apps. Thus, the feedback should warn users about unsafe behavior. For example, if the Google Chrome browser had warned its users about the extension requesting admin-level permissions, they may have decided not to install it.

User Trust

Trust in the digital world is fundamental. The wide adoption of the Internet implies that the public primarily trusts it, its leading providers, and the institutions responsible for keeping its infrastructure up and running. For example, the domain name system (DNS) was exploited in this attack. The DNS is one of the most fundamental components of the Internet, as it's responsible for maintaining records for mapping domain names to IP addresses. If such systems can be targeted, the public's trust in the Internet will wane.

User Knowledge and Awareness

Users also need to educate themselves about the gadgets and technologies they are using daily. A user aware of the risks and implications of cyberattacks is the best defense against any cybersecurity risk. Thus, to solidify the people's trust, more transparency is required in the data flow of the Internet. The protocols and systems governing the information exchange must be re-architected to protect users' privacy and data online.

CONCLUSION

The Google Chrome Extension Spyware is an interesting case of an attack that uses a combination of different factors.

The lack of awareness of how to trust an application source, the attractiveness of free applications, the failure to pay attention to the app's permission request, and the failure to read the terms of usage all contributed to the success of this massive attack.

REFERENCES

1. Joseph Menn J. (2020). "Massive Spying on Users of Google's Chrome Shows New Security Weakness," Reuter, June 17, 2020. https://www.reuters.com/article/us-alphabet-google-chrome-exclusive/exclusive-massive-spying-on-users-of-googles-chrome-shows-new-security-weakness-idUSKBN23P0JO.
2. Company C. (2020). "Google Removes 106 Chrome Extensions for Collecting Sensitive User Data," ZDNet, Jun 18, 2020. https://www.zdnet.com/article/google-removes-106-chrome-extensions-for-collecting-sensitive-user-data/.

3. Awake Security. (2020). “The Internet’s New Arms Dealers: Malicious Domain Registrars,” AWAKE, 2020. https://arista.my.site.com/AristaCommunity/s/article/The-Internet-s-New-Arms-Dealers-Malicious-Domain-Registrars.
4. Osborne C. H. (2021). “Android Spyware Strains Linked to State-Sponsored Confucius Threat Group,” ZDNet, February 11, 2021. https://www.zdnet.com/article/new-novel-android-spyware-linked-to-pro-india-confucius-threat-group/.
5. Moallem A. (2017). Do you really trust “privacy policy” or “terms of use” agreements without reading them? In *Advances in Human Factors in Cybersecurity*. Springer, New York City, pp. 290–295.
6. Dan A. (2008). The Cost of Zero Cost. In *Predictably Irrational*, HarperCollins, New York City, pp. 103–116.

7 Marriott International

Abstract

This chapter after briefly describes the case; examines what we know about it, the data compromised, and the case from a human factors point of view.

Marriott International, Inc., an American multinational company that operates, franchises, and licenses lodging brands that include hotel, residential, and timeshare properties, informed customers in November 2018 of a data breach resulting in the possible disclosure of credit cards, passport numbers, and other personally identifying info belonging to customers [1]. The breach was discovered in November 2018 but has been ongoing since 2014. It involved unauthorized access to the Starwood guest reservation database, which Marriott had acquired in 2016. As a result of this incident, the personal data of millions of Starwood guests was exposed. This occurrence resulted in Marriott incurring significant costs in the recovery process, legal complications, and damage to its reputation [2,3].

After briefly describing the case, this chapter examines what we know about it, the data compromised, and the case from a human factors point of view.

CASE DESCRIPTION

In 2014, two years before the Marriott acquisition, cyberattackers infiltrated Starwood's guest reservation system. Several vulnerabilities in Starwood's system could have contributed to this breach. In September 2016, Marriott officially acquired Starwood. During the acquisition process, Marriott didn't do a detailed cybersecurity audit of Starwood's networks and technology. Some experts suggested that data breaches were related to insecure web applications.

On September 8, 2018, Accenture, a company running IT and information security for Starwood and managing the Starwood Guest Reservation Database, contacted Marriott's IT team and informed them about the Guardium (an IBM security product) alert that was triggered by a query from an administrator's account. The analysis revealed that the person to whom that account was assigned was not the one who made the query; someone else had managed to take control of the account. Following that, investigators began scouring the system for clues and discovered how the breach might have happened.

The breach is believed to have been carried out by a state-sponsored hacker group, exploiting vulnerabilities in Starwood's systems that existed before Marriott's acquisition.

The Marriott case is one of the largest data breaches in history and underscores the importance of strong cybersecurity practices, particularly in mergers and acquisitions.

DOI: 10.1201/9781003461739-9

WHAT WE KNOW ABOUT THIS CASE

According to reports, in 2014, cyberattackers were able to access the guest reservation system of Starwood, via a Remote Access Trojan (RAT). A RAT is malware attackers used to gain full administrative privileges and remote control of a target computer. RATs are often downloaded along with legitimate user-requested programs or are sent to the target as an email attachment via a phishing email. Thus, the RAT gave the cybercriminals unauthorized administrative control over the system, and they were able to exploit digital vulnerabilities, such as outdated versions of Windows Server and open remote desk protocol (RDP) ports, which went unnoticed by Starwood [3].

Then, when Marriott acquired Starwood in 2016, Marriott was unaware that Starwood's workplace devices were infected with malware in 2015 due to a separate attack [4].

Marriott continued to allow Starwood to operate with its compromised system and devices while also migrating customers' sensitive data. Despite the encryption, the cybercriminals decrypted the data and exfiltrated it before re-encrypting it to avoid detection.

It was in September 2018 that Marriott discovered the breach when their internal security tool flashed the alert. Marriot then started an investigation, which led to the public disclosure of the breach.

Marriott then sent emails on November 30, 2018, to affected guests whose email addresses are in the Starwood guest reservation database, offering free credit monitoring for a year [5].

November 30, 2018 [3]. Nearly 500 million customers worldwide, including those in the United States, Canada, and the United Kingdom, had their personal information compromised as a result of the breach.

INFORMATION COMPROMISED

The personal information of nearly 500 million customers worldwide, including those in the United States, Canada, and the United Kingdom, had their personal information compromised because of the breach. The personal information stolen, including name, mailing address, phone number, email address, passport number, Starwood Preferred Guest ("SPG") account information, date of birth, gender, arrival and departure information, reservation date, and communication preferences. For some of these guests, payment card data was also stolen, but Marriott did not say for how many [6].

Even though the debit and credit cards were secured using AES-128 encryption, some of these visitors' payment card information and expiry dates were also disclosed. For the other guests, simply their names and occasionally other details like email addresses and postal addresses were taken [5]. Security experts have expressed concern that cybercriminals could use the stolen information for fraudulent purposes, and that this breach should serve as a warning to all companies to take their customers' data security more seriously.

WHAT CYBERATTACKERS CAN DO WITH HACKED DATA

Data breaches involving the consumer data are usually believed to be done by cybercriminals who are interested in doing credit card frauds or selling the stolen database. However, according to reports from the *New York Times* and the *Washington Post* from December 2018, Chinese intelligence services were linked to the Marriott data breach [7,8].

The New York Times and the *Washington Post* had insider information about the data breach, and they identified that the techniques and code used for this attack were in line with the tactics used by Chinese state-sponsored hackers. The attackers also made use of cloud storage services that Chinese hackers commonly utilize. Moreover, none of the data that was taken appeared on the dark web, which suggested that it was not just a criminal operation.

The purpose of the attack seems to collect extensive data of U.S. government employees and intelligence personnel. The fact that Marriot hotels are the primary stay for the people from the U.S. government and military, and the stolen information can be used to monitor their activity around the world, supports this theory behind the data breach [5].

The Marriott breach seems to be similar to the case of the Office of Personnel Management's (Chapter 9) with the aim of creating a large pool of information on American government employees and agents, which can be analyzed using big data techniques [6].

Looking back, it seems a little odd that during its acquisition of Starwood, Marriott had to fight a takeover effort by the Chinese firm Anbang. It's crucial to remember that the Chinese hackers already had access to Starwood's systems before this, so it might have been a coincidence [4].

In February 2020, the U.S. Department of Justice charged four Chinese military personnel with participating in the 2017 Equifax attack. Sensitive information of millions of people was stolen as a result of this assault. The indictment's statement pointed out that the Marriott and OPM hacks, as well as the Equifax breach, were all components of a larger scheme [9]. To avoid backlash against American operations, the U.S. government typically avoids bringing legal proceedings against foreign officials. This emphasizes the seriousness with which the American government viewed this attack and its gravity [10].

HUMAN FACTORS ANALYSIS

The case illustrates how human error can compromise the security of even the most technologically advanced systems. Marriott's failure to detect the breach for four years and its failure to encrypt passport numbers are examples of human error that contributed to the breach. Several areas, again, are related to the human agents' behavior and vulnerability based on what we know.

MANAGEMENT BEHAVIOR

The first human related issue was the management behavior related to Marriott's slow response to the incident. Marriott discovered the breach in September 2018, but did not inform its customers until November 2018. This delay in response gave the attackers more time to exploit the stolen data, increasing the risk of identity theft and fraud. This suggests that Marriott did not have a robust incident response plan in place or failed to follow the plan properly.

Security and Development Process

The Marriott Data Breach case highlights the importance of designing secure systems. The vulnerability in the Starwood reservation database was related to the web application framework used in the database, which was not properly secured. This suggests that the developers of the application did not prioritize security in the design and development process. Additionally, Marriott's failure to encrypt passport numbers and other sensitive information suggests that the company did not consider the potential consequences of a data breach. This highlights the importance of incorporating privacy and security considerations into the design process of software systems.

Employees Awareness

One of the main human factors involved in the Marriott Data Breach case was a lack of cybersecurity awareness among its employees. The attackers gained access to the database using valid credentials stolen from a Starwood employee. This suggests that the employee was either unaware that their credentials had been stolen or did not follow proper cybersecurity protocols, such as changing their password frequently or using two-factor authentication. Employees should receive training on phishing awareness. Such training helps them avoid compromising their credentials, downloading malicious attachments, or inadvertently sending sensitive information to impersonators.

CONCLUSION

In conclusion, the Marriott Data Breach incident emphasizes how crucial human elements are to cybersecurity. The case illustrates how a combination of technical vulnerabilities and human error can lead to a major data breach with significant consequences for individuals, organizations, and countries. It is essential for companies to prioritize cybersecurity awareness and education among employees, develop robust incident response plans, and incorporate privacy and security considerations into the design and update process of software systems.

REFERENCES

1. Perlroth N., Tsang A., Adam Satariano A. (2028). "Marriott Hacking Exposes Data of Up to 500 Million Guests," New York Times, November 30, 2018. https://www.nytimes.com/2018/11/30/business/marriott-data-breach.html
2. Kelli Young K. (2021). "Cyber Case Study: Marriott Data Breach October 11, 2021," Coverlink Insurance, October 11, 2021. https://coverlink.com/case-study/marriott-data-breach/.

3. U.S. Senate. (2029). "Testimony of Arne Sorenson, President & CEO, Marriott International before the Senate Committee on Homeland Security & Governmental Affairs Permanent Subcommittee on Investigations," hsgac.senate.gov, March 7, 2019. https://nsarchive.gwu.edu/document/20195-national-security-archive-229-testimony-arne.
4. Fruhlinger J. (2020). "Marriott Data Breach FAQ: How Did It Happen and What Was the Impact?" CS, February 12, 2020. https://www.csoonline.com/article/567795/marriott-data-breach-faq-how-did-it-happen-and-what-was-the-impact.html.
5. Ashford W. (2018). "Marriott Data Breach Highlights Basic Failings," Computer Weekly, September 30, 2018. https://www.computerweekly.com/news/252453521/Marriott-data-breach-highlights-basic-failings.
6. Hollander J. (2023). "Marriott Data Breach FAQ: What Really Happened?" Hotel Tech Report, February 16, 2023. https://hoteltechreport.com/news/marriott-data-breach.
7. Timberg C., Nakashima E. (2018). "U.S. Investigators Point to China in Marriott Hack Affecting 500 Million Guests," Washington Post, December 11, 2018.
8. Sanger D. E., Nicole Perlroth N. et al. (2018). "Marriott Data Breach Is Traced to Chinese Hackers as U.S. Readies Crackdown on Beijing," New York Times, December 11, 2018. https://www.nytimes.com/2018/12/11/us/politics/trump-china-trade.html.
9. Fruhlinger J. (2020). "Equifax Data Breach FAQ: What Happened, Who Was Affected, What Was the Impact?" CSO, February 12, 2020. https://www.csoonline.com/article/567833/equifax-data-breach-faq-what-happened-who-was-affected-what-was-the-impact.html.
10. Ting D. (2016)." The Inside Story of Anbang's Final, Failed Attempt to Acquire Starwood," skift.com, April 1, 2016. https://skift.com/2016/04/01/the-inside-story-of-anbangs-final-attempt-to-acquire-starwood/.

Part Three

Sabotage

- Stuxnet
- Colonial Pipelines
- Ukrainian Power Grid Hack

8 Stuxnet

ABSTRACT

This chapter, after a brief description, looks at what we know about the Stuxnet case, and analyzes the case from a human factors point of view.

INTRODUCTION

Stuxnet is a sophisticated computer worm malware initially developed by the U.S. National Security Agency (NSA), the CIA, and Israeli intelligence to sabotage Iran's nuclear facilities. It was first discovered in 2010 and is rumored to have been created and modified since 2005. According to some reports, Stuxnet was built around November 2005; it became known to malware scanners in November 2007 [1,2].

Stuxnet was primarily designed to attack Supervisory Control and Data Acquisition (SCADA) systems, which are technological frameworks used in industrial applications like manufacturing and scientific research. Stuxnet is believed to have played a significant role in damaging and slowing down the Iranian nuclear research and development program, thus serving as a cyberweapon called the "Operation Olympic Games" to stop or hamper the nuclear program of Iran [3].

After a brief description, this chapter looks at what we know about this case and the data compromised, and analyzes the case from a human factors point of view.

CASE DESCRIPTION

Stuxnet was designed to infect networks that are not connected to the Internet. An infected device, such as a USB key, would be sufficient to infect one computer on the target network. Stuxnet would then spread through the network using peer-to-peer (P2P) methods. A P2P is an application architecture that partitions tasks or workloads between peers. Peers are equally privileged, equipotent participants in the network.

Stuxnet malware is highly sophisticated and exploits multiple previously unknown Windows zero-day vulnerabilities to infect computers and spread. Stuxnet's purpose was not just to infect personal computers (PCs) but also to cause real-world physical damage.

The original version of the Stuxnet malware attack targeted the Programmable Logic Controllers (PLCs) used to automate machine processes. For example, it targeted Iranian centrifuges to produce the enriched uranium that powers nuclear weapons and reactors. The worm was designed first to check if that computer was connected to specific models of programmable logic controllers (PLCs) manufactured by Siemens. Then, it altered the PLCs' programming, resulting in the centrifuges being spun too quickly and for too long, damaging or destroying the delicate

DOI: 10.1201/9781003461739-11

equipment. While this happened, the PLCs told the controller computer that everything was working fine, making it difficult to detect or diagnose what was going wrong until it was too late [4].

WHAT WE KNOW ABOUT THIS CASE

Stuxnet explicitly targets PLCs, which allow the automation of electro-mechanical processes such as those used to control machinery and industrial processes, including gas centrifuges for separating nuclear material.

Stuxnet uses flaws and vulnerabilities present in the Windows operating system and software made by Siemens called Step7. Many industries, such as manufacturing, use the Step7 software to program and control PLCs, which are small computers that automate specific repetitive processes in manufacturing and factories.

In 2010, according to reports, the Stuxnet worm damaged Iranian PLCs in the Natanz plant. It collected information on the systems and damaged centrifuges used in the plant. The malware "manipulated valves on the centrifuges to increase the pressure inside them and damage the devices as well as the enrichment process." "Centrifuges are large cylindrical tubes—connected by pipes in a configuration known as a 'cascade'—that spin at supersonic speed to separate isotopes in uranium gas for use in nuclear power plants and weapons" [5]. However, Stuxnet's design and configuration are "not domain-specific." Therefore, it can compromise systems other than its original targets, such as factory assembly lines or power plants. Based on reports, Stuxnet hampered nearly one-fifth of Iranian nuclear centrifuges. Stuxnet, according to some reports, was believed to have destroyed 1,000 centrifuges without impacting anything else [5].

Stuxnet's structure had three components. First, the worm had to be installed on systems that were not connected to the public internet system. Thus, the attacker needed to use, for example, USB flash drives to transport and transmit to the target machines [1,2].

Once the system was infected, the worm spread to the network of computers and looked for the Siemens Step7 software. If it doesn't find the software, it was designed to become dormant after applying to at most three systems. However, when it found the Step7 software, the worm activated the rootkit file onto the PLCs, and the Step7 software controlled those PLCs. Once infected, it gave incorrect and harmful commands to the PLCs and returned typical data to the user to hide [6].

INFORMATION COMPROMISED

Stuxnet, if installed on systems other than its target, does very little harm, unlike most other malware programs. The objective of Stuxnet is to reach its target and make itself inert in Siemens software.

"The attackers made sure that only their designated targets were hit. It makes itself inert if Siemens software is not found on infected computers and contains safeguards to prevent each infected computer from spreading the worm to more than three others, and to erase itself on June 24, 2012" [4].

For its targets, Stuxnet contains, among other things, code for a man-in-the-middle attack that fakes industrial process control sensor signals, so an infected system does not shut down due to detected abnormal behavior. Such complexity is very unusual for malware. The worm consists of a layered attack against three different systems:

1. The Windows operating system
2. Siemens PCS 7, WinCC, and STEP7 industrial software applications that run on Windows
3. One or more Siemens S7 PLCs [4]

WHAT DID CYBERATTACKERS DO WITH THIS ATTACK?

Since the Stuxnet worm had a clear motive based on its targets and its means of cyberattack, it is clear that it wasn't intended as anything more than a weapon to destabilize and hamper Iran's nuclear program. It was carefully designed not to cause any harm or spy on any other unrelated people. Even today, the entirety of the Stuxnet source program is not revealed or identified.

The Institute for Science and International Security (ISIS) suggests, in a report published in December 2010, that Stuxnet is a reasonable explanation for the apparent damage at Natanz and may have destroyed up to 1,000 centrifuges (10%) sometime between November 2009 and late January 2010 [7].

In this case, the objective seems not to collect data but to use the malware as a cyberwar to sabotage a sovereign country's nuclear facilities. Consequently, it does not seem that any data was compromised.

HUMAN FACTORS ANALYSIS

The Stuxnet case is a complex attack case. It included three phases: designing malware (not in the scope of this book), inserting it in one or several target computers, and tricking the supervisory agents once the malware becomes active.

Inserting Malware

The Stuxnet attack was backed by a few major countries' resources and motivation to make it a success since it had a significant political impact. The computer systems in the Iranian nuclear Natanz facility are highly secured and not connected to the public internet system. Consequently, the attackers most likely placed the worm on the computer of an individual who directly worked in the Natanz facility.

It is assumed that the worm was designed to be transported on USB flash drives. Individuals who used infected USB flash drives outside the Natanz facility and connected them to the Natanz systems transmitted the worm to the target.

To get their weapon into the plant, the attackers used four companies that had some connection to Natanz as contractors. They provided a gateway to pass Stuxnet to Natanz through infected employees [8].

Thus, it is safe to assume that the spy agencies' must have been involved in finding an individual with access to the nuclear plant and finding the time to place the malware on one or multiple vulnerable users' computers. The attack's success was primarily related to human agents who might have left their computers unlocked even for a short period and allowed attackers to place the malware. Even for a brief moment, not following security policy comes at a high price, particularly in vital areas.

Tricking Supervisory Agents

The Stuxnet worm took advantage of flaws in the Siemens Step7 software, which is used to program and manage ICS components. It remained hidden until specific conditions triggered it. Thus, it evaded detection by the supervisory agents and other security measures.

The first outsiders to notice the effects of the worm were inspectors from the International Atomic Energy Agency (IAEA). They were permitted access to the Natanz facility, and an unusually high number of damaged centrifuges (around 2,000) were found to be inoperable. However, nobody knew that computer malware was causing damage [9]. The next event was when Stuxnet spread beyond the Natanz facility. One of the possibilities was that it had escaped due to poor security practices by the Iranians at Natanz, something like someone taking a work laptop home and connecting it to the Internet. Cyber experts also concluded that the Stuxnet code was too sophisticated to be the work of amateur hackers, and it needed a team of a minimum of ten people at least two or three years to create it [10].

The active malware periodically modified the output frequency, sabotaging the automation system. Furthermore, Stuxnet replayed old, normal values to the visual SCADA components of the Simatic system so that operators would not suspect a problem. Consequently, the malware might have tricked agents before they found the real issues. Stuxnet's ability to manipulate the ICS and remain hidden from supervisory agents was a key factor in its success. The virus was able to cause significant damage to Iran's nuclear program before it was detected and contained. Once Stuxnet had infected the ICS, it could manipulate the data being fed to supervisory agents [11].

CONCLUSION

The Stuxnet attack in 2010 targeted supervisory control and data acquisition (SCADA) systems to damage Iran's nuclear program despite the technical sophistication relying on vulnerable human behavior [12]. The Stuxnet case showed that weaponized computer malware would compromise human safety since the effect of such attacks not only compromises industrial networks but all human beings.

REFERENCES

1. Fruhlinger J. (2022). "What Is Stuxnet, Who Created It, and How Does It Work?" CSO, August 22, 2017.
2. Fruhlinger J. (2022). "Stuxnet Explained: The First Known Cyberweapon," CSOUS, August 31, 2022. https://www.csoonline.com/article/3218104/stuxnet-explained-the-first-known-cyberweapon.html.
3. Kushner D. (2013). The Real Story of Stuxnet. *IEEE Spectrum*. Vol. 50, pp. 48–53.

4. Knell Y. (2022). "Iran Nuclear Programme: Threat of Israeli Strike Grows," BBC, November 23, 2021. https://www.bbc.com/news/world-middle-east-59322152.
5. Vigliarolo B. (2017). "Stuxnet: The Smart Person's Guide," TechRepublic, August 15, 2017. https://www.techrepublic.com/article/stuxnet-the-smart-persons-guide/.
6. Zetter K. (2022). "An Unprecedented Look at Stuxnet, the World's First Digital Weapon," Wired, November 3, 2014. https://www.wired.com/2014/11/countdown-to-zero-day-stuxnet/.
7. Albright D. et al. (2010). "Did Stuxnet Take Out 1,000 Centrifuges at the Natanz Enrichment Plant?" Preliminary Assessment, Institute for Science and International Security, December 22, 2010. https://isis-online.org/isis-reports/detail/did-stuxnet-take-out-1000-centrifuges-at-the-natanz-enrichment-plant/.
8. Schneier B. (2010). "The Story behind the Stuxnet Virus," Forbes, October 7, 20210. https://www.forbes.com/2010/10/06/iran-nuclear-computer-technology-security-stuxnet-worm.html?sh=4700636a51e8.
9. Broad W., Markoff J., Sanger D. (2011). "Israeli Test on Worm Called Crucial in Iran Nuclear Delay," New York Times, January 15, 2011. https://www.nytimes.com/2011/01/16/world/middleeast/16stuxnet.html.
10. Zetter K. (2011). "Report: Stuxnet Hit 5 Gateway Targets on Its Way to Iranian Plant," Wired, February 11, 2011. https://www.wired.com/2011/02/stuxnet-five-main-target/.
11. Kushner D. (2013). The Real Story of Stuxnet. *IEEE Spectrum*. Vol. 53, No. 3, p. 48. https://spectrum.ieee.org/the-real-story-of-stuxnet.
12. Kesler B. (2011). The Vulnerability of Nuclear Facilities to Cyber Attack. *Strategic Insights*. Vol. 10, No. 15. https://www.semanticscholar.org/paper/The-Vulnerability-of-Nuclear-Facilities-to-Cyber-Kesler/740214467168d1826b9e6e25e2194dfdfe5b742f#citing-papers

9 Colonial Pipeline

Abstract

This chapter, after a brief description, examines what we know about this Colonial Pipeline case and analyzes it from a human factors point of view.

INTRODUCTION

Colonial Pipeline is an American oil pipeline system that carries gasoline and jet fuel mainly to the Southeastern United States. On May 7, 2021, a cyberattack forced the company to close operations proactively and freeze Information technology (IT) systems after becoming a cyberattack victim. The Colonial Pipeline case was a ransomware attack, and it is believed to be the largest successful cyberattack against oil infrastructure in U.S. history [1,2].

Many government resources had to be dedicated to managing the case, including the FBI, the Federal Energy Regulatory Commission (FERC), the Department of Energy, the Department of Transportation, and the executive office. In addition, the hackers intended to cause as much alarm and disruption as possible to force Colonial into complying with their ransom demand. The DarkSide group (which is responsible for this attack) has claimed that it targets large companies that can afford hefty ransoms [1].

After a brief description, this chapter looks at what we know about this case, the data compromised, and analyzes the case from a human factors point of view.

CASE DESCRIPTION

On May 7, 2021, the Colonial Pipeline's control room received a ransom demanding cryptocurrency payment to restore disabled internal systems. The hackers also stole company data that they threatened to leak if the ransom was not paid.

After receiving the demand, the company decided to proactively shut down all 5,500 miles of the pipeline, taking down the largest fuel pipeline in the country and causing buyer panic and shortages across the East Coast [3,4].

WHAT WE KNOW ABOUT THIS CASE

Colonial Pipeline's Chief Executive told a U.S. Senate committee that the attack occurred using a legacy Virtual Private Network (VPN) system without multi-factor authentication. That means one could access through a password without a second step, such as a text message—a common security safeguard in more recent software or force password change feature that requires users to change their passwords, typically after a time period [5]. As a result, hackers could get into the system by stealing a single password.

 DOI: 10.1201/9781003461739-12

On April 29, 2021, a group of hackers connected to the Russia-linked cybercrime group DarkSide gained entry into Colonial's networks through an inactive VPN account. Although the account was no longer used, the hackers could still access the network after supposedly finding the leaked password on the dark web [5].

About a week later, on May 7, a Colonial employee saw the ransom note in the company's control room at around 5:00 AM, demanding approximately $4.4 million in cryptocurrency. It was also discovered that the company's billing infrastructure had been compromised. The hackers sought ransom payment in exchange for a decryptor tool (to decrypt an encrypted file) to restore the system [6,7]. To prevent further damage to its networks, Colonial began shutting down its entire pipeline for the first time in its 57-year history. The Colonial CEO stated, "We had no choice at that point. At that time, we had no idea who was attacking us or what their motives were" [2,3]. By around 6:10 AM, the pipeline was completely shut down. While reports on subsequent case details vary, Colonial CEO later testified that Colonial paid the $5 million ransom the following day in exchange for the decryptor. However, the decryptor tool was so inefficient that Colonial relied more on its backup tools to help restore their disabled networks [4].

The shutdown resulted in widespread disruption. Metaphorically called the "jugular of the U.S. pipeline system," the Colonial pipeline delivers almost half of all the gasoline, diesel, jet fuel, and home heating oil used on the East Coast. Gas prices spiked within a matter of days, reaching their highest level in nearly six years in the Southeastern United States [4].

With the pipeline shut down, Colonial had to assess the severity and scope of the attack. Cybersecurity firm Mandiant began sweeping Colonial's networks to understand how far the hackers had gotten into the system. It was essential to evaluate any possible risks to Colonial's operational technology, which controls fuel flow within the pipeline.

Mandiant concluded that hackers had accessed the company's informational technology network but did not breach the operational systems [2].

Once they were confident that the threat had been contained, Colonial began resuming service on May 12, five days after receiving the ransom note. During its network sweep, Mandiant installed detection tools to help fortify the company's systems. Colonial also examined the entire pipeline after the cyberattack and did not find signs of physical damage. Later, the Justice Department reported that it had recovered approximately $2.3 million of the cryptocurrency ransom, with the discrepancy due to the falling value of bitcoin at the time [5,10]. No repeat threats have been reported since the initial attack.

INFORMATION COMPROMISED

The hackers primarily attacked the company's IT systems. The attackers compromised several finance systems critical to Colonial. They stole about 100 gigabytes of data containing the personal information of about 6,000 individuals, primarily current or former Colonial employees and their families. This information included "names, contact information, birth dates, Social Security, driver's license and military I.D. numbers, and health insurance information" [6,9]. The hackers reportedly threatened to leak this information online if Colonial did not meet the attackers' ransom demands.

During the attack, supply shortage concerns prompted gasoline futures to reach their highest level in three years. As a result, demand has risen, but drivers were urged not to panic. However, the prices increased due to the pipeline disruption [1].

WHAT CAN CYBERATTACKERS DO WITH HACKED DATA?

The hackers seemed not to intend to use Colonial data directly but to cause enough disruption to spur widespread panic and outages. That said, the attackers still use the hacked data for many other types of attacks, such as phishing and social engineering, in order to trick employees [9].

Regional emergencies were declared for 17 affected states. As gas stations ran out of fuel and prices soared, people started panic buying and hoarding, thus worsening the shortage. Airline operations were also disrupted [10].

In addition, such a large-scale attack posed severe challenges for recovery. For example, following the attack, Colonial CEO stated that it would cost Colonial "tens of millions of dollars" to fully restore its systems over the next several months [8]. This cost added to the millions already accumulated with the ransom payment and any costs associated with the week-long pipeline shutdown.

HUMAN FACTORS ANALYSIS

Several human factors-related elements contributed to the Colonial Pipeline attack's success. The review of available reports suggests that the main elements were weak passwords, poor system patching practices, and inadequate employee training and awareness.

The Colonial Pipeline case also required that government and infrastructure organization agents review their procedures for protecting infrastructure. Some experts called this attack the "cybersecurity's Pearl Harbor moment" [11].

PASSWORD

From a human factors point of view, in the Colonial Pipeline case, the attack's initial entry point was a compromised password that allowed the attackers to access the company's VPN. Thus, a single password was enough to enter Colonial systems. Moreover, the CEO admitted that the hackers had exploited an old account not protected by multi-factor authentication, which is one of the crucial and most recommended methods for securing accounts [2,6].

This attack demonstrates that vulnerabilities were related to the lack of security of the administrative policy. Organizations need to have protocols in place to address vulnerabilities in their existing systems and anywhere in their networks that could serve as entry points for hackers, even if they are outdated. These cybersecurity processes should be ongoing and continuously updated.

Cybersecurity Processes

In addition to system weaknesses, federal officials reportedly criticized Colonial for not immediately involving the Cybersecurity and Infrastructure Security Agency (CISA) in post-attack investigations. Inter-organizational complications highlight the ineffective collaboration between private and public agencies in managing large-scale cyberattacks. Consequently, stricter cyber-hygiene standards must be in place to avoid such attack-related shortcomings by not having cybersecurity processes. In this case, it needed protocols to address vulnerabilities in its existing systems and anywhere in its networks that could serve as entry points for hackers, even if they were outdated. These cybersecurity processes should be ongoing and continuously updated [7].

Weak cyber-hygiene standards make it difficult for these systems to recover efficiently and put them (and other sectors) at risk for repeat threats. Several experts have called for stricter cyber hygiene standards and clear-cut protocols for reporting and post-attack investigations. The Colonial case, in particular, calls attention to the cybersecurity standards at privately held companies. Private companies are free to implement whatever security measures they feel are sufficient. The very public impact of the Colonial case may strengthen demands for private companies to exercise better cybersecurity practices [12,13].

Managers and Employees Training and Awareness

One of the important factors in the Colonial Pipeline is inadequate training and awareness.

The employees, IT, managers, and users needed to be sufficiently trained on how to detect and respond to the attack since the attackers had sufficient time to infiltrate deep into the Colonial Pipeline system and cause more significant damage. For example, the inactive accounts should have been closed or at least needed a special procedure to be re-activated.

Decision-Making

Colonial Pipeline's decision to pay the ransom may have been influenced by the pressure to restore the pipeline's operations quickly and the fear of potential legal and reputational consequences. This underscores the need for organizations to have well-defined incident response plans that address the ethical and legal considerations of paying ransoms.

CONCLUSION

Ransomware is becoming increasingly common and poses a serious threat to critical infrastructure. The appeal of large payouts and the unregulated growth of cryptocurrency have contributed to this increase [6]. Attacks like the Colonial Pipeline case are likely to occur again, and based on existing trends, they are forecasted to be even more damaging. These systems must prioritize cybersecurity and the importance of human agent vulnerability from administrative control to protect against future threats and prevent potentially crippling disruptions.

Overall, the Colonial Pipeline ransomware attack underscores the critical role that human factors play in cybersecurity incidents. The incident provides a case study for the analysis of human factors in security incidents and highlights the need for a multidisciplinary approach to addressing cybersecurity challenges. Organizations must prioritize training and awareness programs to prevent similar attacks, implement strong password policies and multi-factor authentication, and ensure timely and thorough system patching practices.

REFERENCES

1. Gonzalez G. et al. (2021). "'Jugular' of the U.S. Fuel Pipeline System Shuts Down after Cyberattack," Politico, May 8, 2021. https://www.bloomberg.com/news/articles/2021-06-04/hackers-breached-colonial-pipeline-using-compromised-password#xj4y7vzkg.
2. Osborne Ch. (2021). "Colonial Pipeline Attack: Everything You Need to Know," ZDNET, May 13, 2021. https://www.zdnet.com/article/colonial-pipeline-ransomware-attack-everything-you-need-to-know/.
3. Turton W., Mehrotra K. (2021). "Hackers Breached Colonial Pipeline Using Compromised Password," Bloomberg, June 4, 2021. https://www.bloomberg.com/news/articles/2021-06-04/hackersbreached-.
4. The Stack. (2021). "Colonial Pipeline "paid $5m ransom" but Decryptor Didn't Work," The Stack, May 13, 2021. https://thestack.technology/colonial-pipeline-paid-5m-ransom/.
5. Target E. (2021). "The Pipeline Hack: 2 (Just 2!) Key Takeaways," The Stack, May 12, 2021. https://www.thestack.technology/pipeline-hack-lessons-two/.
6. Kelly S., Resnick-ault J. (2021). "One Password Allowed Hackers to Disrupt Colonial Pipeline, CEO Tells Senator," Reuters, June 9, 2021. https://www.reuters.com/business/colonial-pipeline-ceo-tells-senatecyber-defenses-were-compromised-ahead-hack-2021-06-08/.
7. Vanessa Romo V. (2021). "How a New Team of Feds Hacked the Hackers and Got Colonial Pipeline's Ransom Back," NPR, June 8, 2021. https://www.npr.org/2021/06/08/1004223000/how-a-new-team-of-feds-hacked-the-hackers-and-got-colonial-pipelines-bitcoin-bac.
8. Morrisonsara S. (2021). "How a Major Oil Pipeline Got Held for Ransom," VOX, June 8, 2021. https://www.vox.com/recode/22428774/ransomeware-pipeline-colonial-darkside-gas-prices.
9. Fung B. (2021). "Colonial Pipeline Says Ransomware Attack Also Led to Personal Information Being Stolen," CNN, August 16, 2021. https://www.cnn.com/2021/08/16/tech/colonial-pipeline-ransomware.
10. Morrison S. (2021). "How a major oil pipeline got held for ransom Threat Remains," Vox. Jun 8, 2021. https://www.vox.com/recode/22428774/ransomeware-pipeline-colonial-darkside-gas-prices.
11. ReederJoe R., McQuad, P. F. (2021). "Cybersecurity's Pearl Harbor Moment: Lessons Learned from the Colonial Pipeline Ransomware Attack," GT Law, Summer 2021. https://www.gtlaw.com/en/insights/2021/8/published-articles/cybersecuritys-pearl-harbor-moment.
12. Sanger D. E., Perlroth N. (2021). "Pipeline Attack Yields Urgent Lessons about U.S. Cybersecurity," New York Times, May 14, 2021. https://www.nytimes.com/2021/05/14/us/politics/pipeline-hack.html.
13. Committee on Homeland Security. (2021). "Cyberthreats in the Pipeline: Lessons from the Federal Response to the Colonial Pipeline Ransomware Attack," US Congress. https://www.congress.gov/event/117th-congress/houseevent/LC67088/text.

10 Ukrainian Power Grid Hack

Abstract

This chapter, after a brief description of the Ukrainian Power Grid hack, looks at what we know about this case, the data compromised, and analyzes the case from a human factors point of view.

INTRODUCTION

On December 23, 2015, Ukraine's power grid was hacked, resulting in power outages for roughly 230,000 consumers in Ukraine for one to six hours. The attack took place during a Russian military 2014 intervention in Ukraine. It was attributed to a Russian advanced persistent threat group known as "Sandworm." Sandworm Team is a known sophisticated group that seems to be affiliated with the Russian government [1]. Espionage and sabotage operations are hard to investigate because, contrary to attacks for financial gain, credit cards or bank accounts are not trackable or monetizable on black markets [2].

After a brief description of the case, this chapter looks at what we know about this case, the data compromised, and analyzes the case from a human factors point of view.

CASE DESCRIPTION

According to some reports, an employee of the power grid noticed that the cursor on his computer began to move around the screen independently. He tried to gain control of his computer, but it was too late. He had been logged out and 225,000 residents were left in the dark. The attackers carefully planned their assault over many months, first doing reconnaissance to study the networks and siphon operator credentials, then launching a synchronized assault in a well-choreographed dance [1].

WHAT WE KNOW ABOUT THIS CASE

Much earlier in 2015, through a spear-phishing campaign, attackers targeted the IT staff and system admins working for various electricity distribution companies in Ukraine. The spear-fishing campaign delivered targeted employees a malicious email that, and when one clicked on its attachment, a popup would appear asking the user to enable macros for the document. Enabling the macros resulted in a program called BlackEnergy3 infecting their machines. Attackers then conducted extensive reconnaissance for several months, which finally resulted in gaining access to the

DOI: 10.1201/9781003461739-13

Windows Domain Controllers, where user accounts for networks are managed. They used worker credentials logged in to the Supervisory Control and Data Acquisition (SCADA) network. Once the attackers broke into the SCADA networks, they finalized their attack on December 23 when the employees at the control center realized someone on the outside had gained control. At the same time, the attackers launched a telephone denial-of-service attack (TDoS). TDoS is accomplished when an attacker successfully consumes all available telephone resources so that there is no unoccupied telephone line against customer call centers to prevent customers from calling in to report the outage [3,4].

INFORMATION COMPROMISED

The attackers aimed to sabotage the grid system. However, once they got control of the distribution management system of the grid and studied each of them carefully, they were able to write malicious firmware to replace the legitimate firmware. There is no report of any data breach. However, once attackers were in the system, they could have obtained as much as they wanted. Because the data of information from spring or cyberwar is generally not sold on the Dark Web, it is hard to know what was taken.

WHAT CYBER-ATTACKERS CAN DO WITH HACKED DATA?

Ukrainian grid hacks have served Russian interests in at least two ways. First, they are a component of a destabilization campaign aimed at Ukraine as it reduces its dependence on Russia and leans west toward the European Union (EU) and NATO economically, politically, and militarily. Second, the attacks were meant to demonstrate the offensive capabilities of Russian hackers and allowed Russia to prove its effectiveness on a country that cannot retaliate in kind.

HUMAN FACTORS ANALYSIS

An E-ISAC analytic report summarizes the technical components used by the attackers of the Cyberattack on the Ukrainian Power Grid.

- Spear-phishing to gain access to the business networks of the power suppliers.
- Identification of BlackEnergy 3 at each of the impacted power suppliers.
- Theft of credentials from the business networks.
- The use of VPNs to enter the ICS network.
- The use of existing remote access tools within the environment or issuing commands directly
- from a remote station similar to an operator HMI.
- Serial-to-ethernet communication devices impacted at a firmware level16.
- The use of a modified KillDisk to erase the master boot record of impacted organization systems as well as the targeted deletion of some logs17.

- Utilizing UPS systems to impact connected load with a scheduled service outage.
- TDoS attack on the call center [5].

With all its complexity, the cause of the Ukrainian power grid attack was again human vulnerability. In the spear-phishing campaign targeting IT staff and system admins, clicking on an email attachment was sufficient to open a backdoor for the attackers. Attackers then installed malware in a few minutes, took remote control of the human–machine interface (HMI), and switched off most of the switchgear of the grid. The reaction time of the attackers was much faster than that of the Ukrainian power IT teams. It is also reported that attackers used a custom-developed exploit to prevent the operator from regaining control of the network by wiping out many disks (using KillDisk) and overwriting the Ethernet-to-serial gateway firmware with random code, thus turning the devices into unrecoverable pieces of scrap. In contrast, the Ukrainian power grid operators did not have a rescue plan, thus indicating another human-related failure.

It is also reported that the operators, due to the high traffic volume, could not detect the signs of cyberattacks. The reason could be related to how the traffic volume applications were designed. A human operator might be unable to identify the few suspect packets in the high-volume traffic [6].

CONCLUSION

The Ukrainian power grid attack on the Colonial Pipeline (Chapter 8) and Stuxnet (Chapter 7) shows how easy it can be to sabotage countries' infrastructures. Attacking the infrastructures of the country can now be considered an important cyberweapon in wars, despite the results of some studies that suggest these types of cyberattacks have had little or no impact on fighting [7,8]. In addition, well-designed defense-in-depth protection effectively addresses the threats facing their organizations [9–12].

Besides the technical solutions to protect the network, preparing human resources for such attacks is as important as the technical solution. Awareness of operators, privilege management, and reaction time and preparedness are fundamental factors in protecting infrastructure facilities.

REFERENCES

1. Zetter K. (2014). "Russian 'Sandworm' Hack Has Been Spying on Foreign Governments for Years," Wired. October 14, 2014. https://www.wired.com/2014/10/russian-sandworm-hack-isight/.
2. Park D., Walstrom M. (2017). "Cyberattack on Critical Infrastructure: Russia and the Ukrainian Power Grid Attacks," The Henry M. Jackson School of International Studies, University of Washington, October 11, 2017. https://jsis.washington.edu/news/cyberattack-critical-infrastructure-russia-ukrainian-power-grid-attacks/.
3. Zetter K. (2016). "Inside the Cunning, Unprecedented Hack of Ukraine's Power Grid," Wired, March 3, 2016. https://www.wired.com/2016/03/inside-cunning-unprecedented-hack-ukraines-power-grid/.

4. Amy Krigman A. (2022). "Cyber Autopsy Series: Ukrainian Power Grid Attack Makes History," globalsign.com, Blog, October 22, 2020. https://www.globalsign.com/en/blog/cyber-autopsy-series-ukranian-power-grid-attack-makes-history.
5. Lee R. M., Assante M. J., Conway T. (2016). "Analysis of the Cyber Attack on the Ukrainian Power Grid. Defense Use Case (PDF)," E-ISAC, March 18, 2016. https://nsarchive.gwu.edu/sites/default/files/documents/3891751/SANS-and-Electricity-Information-Sharing-and.pdf.
6. Patrice Bock P. et al. (2023). "Lessons Learned from a Forensic Analysis of the Ukrainian Power Grid Cyber-Attack," blog.isa.orG, Viewed on September 4, 2023. https://blog.isa.org/lessons-learned-forensic-analysis-ukrainian-power-grid-cyberattack-malwae.
7. Kostyuk N., Zhukov Y. M. (2019). Invisible Digital Front: Can Cyber Attacks Shape Battlefield Events? *Journal of Conflict Resolution*. Vol. 63, No. 2, pp. 317–347. https://doi.org/10.1177/0022002717737138. ISSN 0022–0027. S2CID 44364372.
8. Zetter K. (2016). "Everything We Know about Ukraine's Power Plant Hack," Wired, January 28, 2016. https://www.wired.com/2016/01/everything-we-know-about-ukraines-power-plant-hack/.
9. Beach-Westmoreland N., Styczynski J., Stables S. (2016). "When the Lights Went Out," Booz Allen Hamilton,.boozallen.com/, Booz Allen Hamilton Inc. https://www.boozallen.com/content/dam/boozallen/documents/2016/09/ukraine-report-when-the-lights-went-out.pdf.
10. Sanger D. E., Barnes J. E. (2021). U.S. and Britain Help Ukraine Prepare for Potential Russian Cyberassault. The New York Times. ISSN 0362-4331. Archived from the original on 2022-01-16. Retrieved 2022-01-17.
11. Gamliel A. N. (2017). "Securing Smart Grid and Advanced Metering Infrastructure." https://www.linkedin.com/pulse/securing-smart-grid-advanced-metering-infrastructure-adi-nae-gamliel/.
12. Greenberg A. (2017). How an Entire Nation Became Russia's Test Lab for Cyberwar," Wired, June 28, 2017. https://www.wired.com/story/russian-hackers-attack-ukraine/.

Part Four

Fake News Shaping People's Opinions

- Facebook–Cambridge Analytica
- Instagram

11 Facebook–Cambridge Analytica

Abstract

This chapter, after a brief description of the Facebook–Cambridge Analytica case, examines what we know about it and analyzes it from a human factors point of view.

INTRODUCTION

In March 2018, *The Guardian* reported that Cambridge Analytica Ltd., a British political consulting firm, misused data obtained from Facebook to build voter profiles [1]. According to the *New York Times*, in 2014, contractors and employees of Cambridge Analytica, eager to sell psychological profiles of American voters to political campaigns, acquired the private Facebook data of tens of millions of users. This case is now considered the most significant known leak in Facebook history. The extent of this case was more than data usage, considering its influence on several elections in the United States and abroad [2]. According to the BBC, the UK News Broadcasting, and a Channel 4 investigation, an executive said Cambridge Analytica and its parent company Strategic Communications Laboratories (SCL) had worked in more than 200 elections across the world, including India, Nigeria, Brazil, Mexico, and the Czech Republic, to name a few.

CASE DESCRIPTION

Cambridge Analytica was a company that provided data-driven solutions and data analytics solutions for businesses. It harvested personal information on where users lived and what pages they liked, which helped build psychological profiles that analyzed characteristics and personality traits. Aleksandr Kogan, a Moldovan-born American data scientist professor at the University of Cambridge, is known for developing an app based on psychological profiles that analyzed characteristics and personality traits. The data came from a personality quiz, which approximately 270,000 people were paid to take. The quiz—"thisisyourdigitallife"—also pulled data from their friends' profiles, resulting in an enormous data stash.

Aleksandr Kogan shared the information in a commercial partnership with Strategic Communication Laboratories (SCL), which later created Cambridge Analytica. The apps allowed Cambridge Analytica to collect the personal details of 80 million Facebook users [1,3].

In March 2018, a Cambridge Analytica employee named Christopher Wylie revealed the firm's activities in interviews with *The Guardian* and *The New York Times*.

DOI: 10.1201/9781003461739-15

In June 2016, Donal Trump's presidential campaign hired Cambridge Analytica to help target ads using voter data gathered from millions of adults in the United States [4].

Data-driven solutions work might benefit many companies or organizations, governments, etc. The problem was, in this case, that the data was acquired by the illegal acquisition of Facebook data of millions of users used by Cambridge Analytica to offer data-driven solutions to their customers. During the presidential election campaign of 2016, the company worked to get the data of users registered as voters in the United States through Facebook even though Facebook users had not authorized the use of this data. Cambridge Analytica used the data for targeted advertising to identify potential voters and customized messages to advertise to Facebook users [2].

WHAT WE KNOW ABOUT THIS CASE

Before May 2015, Facebook had a security vulnerability where access to a user's profile would also give access to that users' friends' data [5]. Upon learning of the vulnerability, Facebook fixed the issue, but they did not inform users whose data may have been compromised due to the vulnerability. Facebook could have performed an analysis as to who accessed the data but chose not to.

In 2013, Cambridge University's academics developed a personality test to understand the psychographic profile of users given a test. The test involved a series of questions that users would answer and use the big five personality traits, often called OCEAN: Openness, Conscientiousness, Extroversion, Agreeableness, and Neuroticism. OCEAN's psychological profile was evaluated, and researchers found that the results reflected the user's overall Facebook profile and activity. However, these researchers refused to share their data and the results obtained. After knowing that the results of personality tests can be linked to the overall Facebook activity of users, a company called Global Science Research (GSR) partnered with Cambridge Analytica to undertake a research project [6].

The research aimed to find parameters for developing users' OCEAN profiles through a personality test. The personality test was launched as a Facebook app, "This is Your Digital Life." The app monitored the profiles of users and their friends. As a result, 87 million Facebook users' data was accessed through 270,000 users using "This is Your Digital Life."

The data was also combined with other publicly available data on social media, online shopping, and criminal records. The data analyzed was used to create 5000+ data points to favor Donald Trump's presidential campaign. The results were used to micro-target users to manipulate their opinions and influence them to vote for the candidate. In a statement to *The Guardian*, the director of King's College's Centre for the Study of Media, Communication, and Power, 40,000–50,000 variant advertisements were posted to users, and the advertisements were adapted according to the response received. The use of Artificial Intelligence (AI) in elections by invading citizens' privacy raised concerns about the fundamental human rights of citizens [7,8].

INFORMATION COMPROMISED

The data was obtained through a personality quiz app called "This Is Your Digital Life," which collected information from users who took the quiz and their Facebook friends, significantly expanding the data pool without explicit consent. This data was then allegedly used to influence voter behavior in political campaigns. The compromised user information included:

Profile Information: This included details such as names, gender, birthdates, locations, and other personal identifiers available on user profiles.

Friend Networks: Data on users' friends and connections were accessed, allowing the collection of information about a wide network of individuals beyond those who directly interacted with the app.

Likes and Interests: Information on pages and posts that users had liked, as well as their interests and preferences, was harvested. This data was used to build detailed personality profiles.

Behavioral Data: This encompassed users' activities on Facebook, including status updates, shared posts, and comments, providing insights into their behaviors and opinions.

Psychographic Data: By analyzing the gathered data, Cambridge Analytica created psychographic profiles, which included inferred psychological traits such as personality types, political leanings, and susceptibility to specific types of messaging [9].

WHAT CAN CYBERATTACKERS DO WITH HACKED DATA?

The information collected from this hack was combined with other publicly available data of users. The data together was used to create psychographic profiles of users. The psychographic profile, along with the likes/dislikes of the person, could reveal the entire personality of the user. The known personality was used to micro-target users with the advertisements they would most likely respond to. The users' response was also given to the AI model to monitor the response and adapt the advertisement. It is not clear how much this service was sold or would be sold. When Mark Zuckerberg, Chief Executive Officer of Facebook, was asked at a Senate hearing about what other firms acquired Aleksandr Kogan's app based on the psychological profiles of Facebook data, he couldn't answer [10,11].

HUMAN FACTORS ANALYSIS

Since this case was revealed, a great deal of research has been conducted from different perspectives and, most importantly, studied the effect on countries' political systems, the democratic election process, shaping public opinion to sell products, etc.

In this case, the human factors perspective can be summarized in the following areas: users' awareness of self-disclosure, mass data analysis of social media user data to shape people's opinions, users' trust in social media, and reliability of the information received through social media.

Users' Awareness of Self-Disclosure

With the rise in internet services, many new websites establish themselves as companies providing service to the user. Examples of such companies are Facebook, Instacart, and Amazon. The sign-up process for these websites is simple. By combining the sign-up process with Google, the sign-up process becomes even easier. Users give many personal, proprietary, confidential details simply trusting the websites. Creating an account generally requires providing an email address, cellphone number, date of birth, location, and more. With some of this information, the other information can be matched from what is available on the internet, and combining all this information gives collectors a mine of information.

Mass Data Analysis from Social Media

All companies have access to large amounts of customer data, but customer privacy is not often considered or prioritized. Similarly, AI tools do not take into consideration users' rights and confidentiality. Furthermore, companies' security vulnerabilities pose another threat to users' data collected and then compromised. Thus, using compromised data with AI tools can be a serious threat to basic human rights. This was the case with the Cambridge Analytica Facebook data breach. Cambridge Analytica used Facebook's Open Graph Interface (API) vulnerability to gather data from 87 million users. Cambridge Analytica further analyzed the data with additional data sources to get the psychographic profiles of the users. The case study explains how the data was used in political campaigns to influence voters.

Users' Trust in Social Media

Data breaches exploit the trust of users. Betrayal of trust affects user trust in each platform for some time. Eventually, users return to the platform unless something better is available.

A good user interface and projections of the number of users already using the website establish trust between the user and the website. The users' trust in the system makes them give unnecessary but important information such as personal information. By combining the sign-up procedure, websites receive user information from partnered websites. For example, when opening an email account, a user might provide details such as a contact recovery email address. When the same email account gets linked to a shopping website, the shopping website also receives the information that is not required from the email account.

If sensitive information is compromised, the organization can also suffer a loss of trust among stakeholders and financial losses. Social media websites can also compromise users' data by sharing it with other companies. Other companies may use the data to gain insights about the users. The Facebook–Cambridge Analytica data breach is a clear example of this.

Social media platforms are widespread, and users very frequently use internet websites. However, the laws such as European Union Data Protection Regulation

(GDPR) around such events are still evolving, so the lack of guidelines from authorities leaves cybersecurity as a severe threat.

Moreover, software engineers are taught to code faster without paying attention to the security details of the code. This results in vulnerabilities like Facebook's Open Graph Interface (API) vulnerabilities. Safety and ethics are treated as add-ons in software designs. They need to be the foundations of the design [9].

Policy development should be based on addressing potential issues with AI technologies. For instance, the Moral Responsibility for Computing Artifacts outlines five rules established by a community of scholars. This is particularly relevant for companies like Cambridge Analytica and Facebook [10].

CONCLUSION

Among all cyberattack cases, the Facebook–Cambridge Analytica incident stands out as one that could have enduring implications for human society and social structures. With the proliferation of AI and social media, concerns about privacy are deservedly heightened. The Facebook–Cambridge Analytica scandal also underlines the vulnerabilities of democratic processes to manipulation through social media and data exploitation. It underscores the need for robust safeguards to protect privacy, promote transparency, and preserve the integrity of democratic elections and public discourse.

REFERENCES

1. Carole Cadwalladr C., Graham-Harrison E. (2018). "Revealed: 50 Million Facebook Profiles Harvested for Cambridge Analytica in Major Data Breach," The Guardian, March 17, 2018. https://www.theguardian.com/news/2018/mar/17/cambridge-analytica-facebook-influence-us-election.
2. Confessore N. (2018). "Cambridge Analytica and Facebook: The Scandal and the Fallout So Far," New York Times, April 4, 2018. https://www.nytimes.com/2018/04/04/us/politics/cambridge-analytica-scandal-fallout.html.
3. Hern A. (2018). "Cambridge Analytica: How Did It Turn Clicks into Votes?" The Guardian, May 6, 2018. https://www.theguardian.com/news/2018/may/06/cambridge-analytica-how-turn-clicks-into-votes-christopher-wylie.
4. Alexandra Ma A., Ben Gilbert B.. "Facebook Understood How Dangerous the Trump-Linked Data Firm Cambridge Analytica Could Be Much Earlier Than It Previously Said. Here's Everything That's Happened up Until Now," Business Insider, August 23, 2019. https://www.businessinsider.com/cambridge-analytica-a-guide-to-the-trump-linked-data-firm-that-harvested-50-million-facebook-profiles-2018-3#what-did-cambridge-analytica-do-1.
5. Isaac M., Frenkel S. (2018). "Facebook Security Breach Exposes Accounts of 50 Million User," New York Times, September 28, 2018. https://www.nytimes.com/2018/09/28/technology/facebook-hack-data-breach.html.
6. Isaak J., Hanna M.J. (2018). "User Data Privacy: Facebook, Cambridge Analytica, and Privacy Protection," Computer, 2018. https://ieeexplore.ieee.org/stamp/stamp.jsp?arnumber=8436400.
7. Bloomberg J. (2017). "Does Trump's 'Weaponized AI Propaganda Machine' Hold Water?," Forbes, March 5, 2017. https://www.forbes.com/sites/jasonbloomberg/2017/03/05/does-trumps-weaponized-ai-propaganda-machine-hold-water/?sh=2af9be7f761c.

8. Anderson B. (2017). "The Rise of the Weaponized AI Propaganda Machine," Medium, February 12, 2017. https://medium.com/join-scout/the-rise-of-the-weaponized-ai-propaganda-machine-86dac61668b.
9. Barrett B. (2018). "How to Check if Cambridge Analytica Could Access Your Facebook Data BRIAN," Wired, April 9, 2018. https://www.wired.com/story/did-cambridge-analytica-access-your-facebook-data/.
10. Locklear M. (2018). "Cambridge Analytica is Closing, but It Solves Nothing," Engadget, May 2, 2018. https://www.engadget.com/2018-05-02-cambridge-analytica-closing-solves-nothing.html.
11. Resnick B. (2018). "Cambridge Analytica's Psychographic Microtargeting. What's Bullshit and What's Legit," VOX, March 26, 2018. https://www.vox.com/science-and-health/2018/3/23/17152564/cambridge-analytica-psychographic-microtargeting-what.

12 Instagram

Abstract

This chapter will focus on the October 2018 Instagram influencers case, a marketing tool for Instagram influencers.

INTRODUCTION

The social media marketing industry has been increasing for the last few years. Influencers on social media play an essential role in today's marketing industry. The Instagram influencer marketing industry has also been growing, as have the marketing agencies that handle some of these influencers' accounts. Influencers have also become a significant part of many big brands' marketing strategies. According to some reports, the influencer market grew steadily; nearly 68% of U.S. marketers from companies with 100 or more employees would use influencer marketing in 2021. The influencer and other paid content represented 20% of U.S. agencies and marketing professionals' digital advertising budgets in 2021 [1]. According to a report, the Influencer Marketing Industry is set to grow to approximately $24 Billion by the end of 2024.

Prolific influencers have millions of followers on multiple social media sites and charge a large sum of money to do paid brand promotions. Thus, attackers target influencers with fake branding to gain access and hack their accounts. Hackers can also communicate directly with influencers, pretending to represent major brands. There is no standard negotiation method between brands and influencers, so they might fall into such traps, as sponsored content deals happen beyond Instagram's official advertisement mechanism.

Data breaches happened several times to Instagram in a short period. For example, in August 2018, hundreds of Instagram accounts were controlled by several Russian hackers to edit the profile information and email addresses for unclear purposes. As a result, Instagram received many emails and complaints from users, mainly about the difficulty of restoring their accounts [2].

After this, in October 2018, many Instagram influencers' accounts were stolen and used to promote scamming ads [3].

Again, in May 2019, a marketing tool for Instagram influencers called Chtrbox exposed 49 million Instagram users' personal information caused by an unprotected Amazon Web Services (AWS) server [4]. It was discovered that many influencers' private contact information, including real names, phone numbers, photos, geographic information, email addresses, and followers, was exposed. The stolen accounts of the influencers were used to promote scamming ads, which might have created more victims since those influencers had millions of followers. The leaked user data was traded through the dark web or made public.

DOI: 10.1201/9781003461739-16

Again, in September 2019, millions of real names and phone numbers of Instagram users were exposed due to vulnerabilities. Instagram claimed that it was patched out before any illegal access was discovered, but this is difficult to confirm [5–7].

This chapter will focus on the October 2018 Instagram influencers case, a marketing tool for Instagram influencers [8].

CASE DESCRIPTION

In October 2018, a publicist of one of the top Instagram influencers received an email from a hacker calling himself "Joshua Brooks" stating that he wanted to start a business partnership and was ready to offer $80,000 for each post.

The hacker claimed to have worked with many of the top Instagram celebrities and was able to convince the publicist. The influencer was asked to log in to a third-party Instagram analytics tool, Iconosquare. Such tools are generally used to track an influencer campaign's success. The link sent to the influencer was of a cloned version of the Iconosquare tool set up for phishing. Once the influencer logged into the cloned version, the hacker gained control of the influencer's account. Within minutes after gaining control, the hacker was spamming the account. When they gained control of the account, the hackers changed the bio of the account to "managed by SCL media." They then tried to reach out to various brands to negotiate a deal for sponsored content [8,9].

WHAT WE KNOW ABOUT THIS CASE

The website that the hacker convinced the influencer to log into was a phishing site instead of the original Iconosquare. Therefore, the hacker took control of the influencer's Instagram account, and immediately, the influencer's millions of followers were spammed for a free iPhone offer. In addition, Joshua Brooks stole several YouTubers and Instagram stars' accounts to promote scamming ads.

The influencer fell for the phishing trick because the individual didn't look closely at the URL address. It was discovered that many Instagram users' private information, including photos, geographic information, email addresses, and followers, was leaked.

This type of attack falls in the category of phishing. Generally, there are two types of phishing attacks to gain control of the user account and its data: Ninja Mode and Owning Mode.

In Ninja Mode, the hacker usually does not change or manipulate the user's account name and password or steal the account. Instead, the hacker keeps track of all the user's activities for an extended period in order to try to learn their patterns. Then, the hacker pretends to be the actual user of that account and tries to ask the user's followers about their personal information. As Instagram provides features such as unsend messages and deleting the conversation, the attacker can easily get rid of their tracks.

In Owning Mode, hackers control Instagram accounts. They alter all the details and information, so the user can never regain account control. That account will now totally belong to the hacker. Once in Owning Mode, hackers can carry out various malicious activities, such as stealing sensitive data, installing malware, or disrupting service. Hackers generally try to sell accounts if there are many followers for the account. They try to steal more data by imitating the user of the stolen account.

INFORMATION COMPROMISED

In this case, the cyberattacker exploited a vulnerability in Instagram's API, allowing them to access users' personal information, including email addresses and phone numbers. The breach occurred when hackers exploited a vulnerability in Instagram's API that was intended for businesses to contact individuals. Attackers retrieved the contact information associated with the Instagram accounts of various celebrities, influencers, and other public figures.

WHAT CAN CYBERATTACKERS DO WITH HACKED DATA?

The attackers can use the compromised email addresses and phone numbers to launch phishing attacks. The stolen personal information can also be used for identity theft purposes. The attackers can also use the data for other social engineering tactics to manipulate or deceive the influencer or their associates.

The attackers might also use the stolen data to harass or extort the influencer. It also allows attackers to post unauthorized content, manipulate followers, or carry out further attacks.

Attackers can trade the data through the Dark Web or make it public. With privacy information, hackers can easily steal Instagram accounts to promote scamming ads or perform other illegal activities. This situation can only be worse for influencers because they have many followers, and thus, more people will be affected by the data breach.

Hackers can set up new accounts with the list of phone numbers attached to user accounts. Then, the hackers can attempt to sync contacts by entering known and valid phone numbers. Once the hackers use valid phone numbers, they can link them to the account number and return their real names [10].

HUMAN FACTORS ANALYSIS

In the case of the Instagram influencer's account, the success of this attack can largely be attributed to the susceptibility of the influencer's behaviors to manipulation. The influencer and publicist failed to verify the identity of "Joshua Brooks." In addition, the influencer needed to tell the difference between lconosquare.biz and iconosquare.com. Consequently, influencers became victims of phishing to lose control of their Instagram accounts, and as we know, phishing relies on human factors.

To steal account or private information, attackers tricked the user, who didn't look closely at the URL address. Because the website address and webpage looked similar to the original, the influencer probably needed to know the potential danger of losing the account. The attacker exploited an Instagram vulnerability. However, the emerging influencer marketing industry also requires legal documentation, formalized communication, and negotiation processes. Since much of this occurs outside of Instagram's official advertising channels, it is challenging for the company to prevent scams [8,9,11].

CONCLUSION

The compromise of personal information, especially for high-profile individuals like Instagram influencers, poses significant risks to both the influencers affected and their followers. It underscores the importance of user awareness about social engineering phishing and multi-factor authentication. Consequently, user education can mitigate the impact of such breaches.

REFERENCES

1. Influencer Marketing Stats. (2021). "How Creators Have Impacted Businesses in 2021," Business Insider Intelligence, May 17, 2021. https://www.businessinsider.com/influencer-marketing-important-for-brands-2021-5.
2. Cuthbertson A. (2018). "People's Instagram Accounts Are Being Mysteriously Taken over by Russians, and They Can't Get Them Back," Independent, September 25, 2018. https://www.independent.co.uk/life-style/gadgets-and-tech/news/instagram-hack-accounts-russia-breached-take-over-accounts-how-locked-2018-a8553776.html.
3. Lorenz T. (2018). "How Hackers Are Stealing High-Profile Instagram Accounts," The Atlantic, November 16, 2018. https://www.theatlantic.com/technology/archive/2018/11/hackers-are-stealing-influencer-instagram-accounts-promising-lucrative-brand-deals/575662/.
4. Ikeda S. (2019). "Instagram Breach Exposes Personal Data of 49 Million Users," PO Magazine, June 3, 2019. https://www.cpomagazine.com/cyber-security/instagram-breach-exposes-personal-data-of-49-million-users/.
5. Ikeda S. (2019). "Instagram Data Leak Exposes Account Information Including Full Names and Phone Numbers," CP Magazine, September 23, 2019. https://www.cpomagazine.com/cyber-security/instagram-data-leak-exposes-account-information-including-full-names-and-phone-numbers/.
6. Hoffman Z. (2019). "Instagram Confirms Security Issue Exposed User Accounts and Phone Numbers—Exclusive," Forbes, September 12, 2019. https://www.forbes.com/sites/zakdoffman/2019/09/12/new-instagram-hack-exclusive-facebook-confirms-user-accounts-and-phone-numbers-at-risk/?sh=39a49fff2200.
7. TaylorLorenzT.(2018)."HowHackersAreStealingHigh-ProfileInstagramAccounts,"The Atlantic, November 16, 2018. https://www.theatlantic.com/technology/archive/2018/11/hackers-are-stealing-influencer-instagram-accounts-promising-lucrative-brand-deals/575662/.
8. Lorenz T. (2018). "How Hackers Are Stealing High-Profile Instagram Accounts in the Wild West of "Influencer" Marketing, There Are Few Protections and Plenty of Easy Marks," The Atlantic, November 16, 2018. https://www.theatlantic.com/technology/archive/2018/11/hackers-are-stealing-influencer-instagram-accounts-promising-lucrative-brand-deals/575662/.
9. Giuseppe D., Ferro L. S., Marrella A., Catarci T., Costabile M. F. (2021). Human Factors in Phishing Attacks: A Systematic Literature Review. *ACM Computing Surveys*. Vol. 54, No. 8, Article 173, 35 pages. https://doi.org/10.1145/3469886.

10. Santora J. (2024). "17 Key Influencer Marketing Statistics to Fuel Your Strategy," Influencer Marketing Hub, February 6, 2024. https://influencermarketinghub.com/influencer-marketing-statistics/.
11. Gallo L., Gentile D. et al. (2018). Phishing Attacks. *Computers & Security*. Vol. 139, April 2024, 1–15. https://www.sciencedirect.com/science/article/pii/S0167404823005813.

Part Five

Data Privacy

- Baidu Android Apps
- Zoom
- LinkedIn
- Spotify

13 Baidu Android Apps

Abstract

This chapter will focus on the Baidu Android Apps case that leaked users' personal information to Chinese servers secretively and analyzes it from a human factors point of view.

INTRODUCTION

Baidu, Inc. is a Chinese multinational technology company specializing in Internet-related services, products, and artificial intelligence. Baidu had two Android-based mobile applications: Baidu Search Box and Baidu Maps. These apps quickly gained popularity among the public, creating a record-breaking number of users who downloaded the app to nearly 6 million [1]. However, in October 2020, Palo Alto Network, Inc., an American multinational cybersecurity Unit42 team, discovered a significant flaw in Baidu's two applications. The Palo Alto Network team found that both apps leaked users' personal information to Chinese servers secretively, even when they didn't state permission to access such user information when they downloaded the app on their phones [1]. Following this revelation, Google removed these applications from the Google Play Store.

CASE DESCRIPTION

Baidu initially released Android-based applications, Baidu Search Box and Baidu Maps, in the Popular Republic of China. They were made available in the United States via the Google Play Store in 2020.

Baidu search box is a search engine that is similar to Google Search. Based on the search string the user gives as input, it provides a list of websites containing the given searched string and ranks the pages based on its internal ranking algorithm [2]. Baidu Maps is a normal location-based map on the mobile platform, like Google Maps, wherein one can locate different places and navigate to those locations via the navigation services provided by the application. These were the stated main objectives and intended uses of these apps when released in the U.S. Google Play Store.

Palo Alto Networks used Machine Learning (ML)-based approaches to identify how the leak occurred via mobile applications. As a result, they found that Baidu's Android software development toolkit was pushing all user data from mobile phones to Baidu servers in China in encrypted form. Thus, on confirming the data leak via Baidu's app, Palo Alto Networks immediately informed Google Play Store, and both these apps were removed from the Play Store by Google in October 2020. Furthermore, all users of these apps were directly informed by Google to stop using them immediately and uninstall them from their smartphones, if possible, to avoid being victims of such cyberattacks [3].

DOI: 10.1201/9781003461739-18

WHAT WE KNOW ABOUT THIS CASE

Users must follow a setup installation stage when downloading a mobile application. During an app's setup, the app's developers need to state up-front the list of permissions that the app might require to be used, such as access to the internet, access to the camera for photo apps, and access to videos for video editor apps and such [4].

In this case, both the apps that Baidu released on the Google Play Store, namely Baidu Search Box and Baidu Maps, didn't request permission from the user regarding access to their personal information like International Mobile Equipment Identity (IMEI), International mobile subscriber identity (IMSI), mobile-model, a media access control address (MAC address), and many more such personal information.

Once the user downloads the application on their mobile phones, Baidu pushes the Software Development Kit (SDK) to get all the information mentioned above from the user's smartphone without informing the user. Baidu then encrypted the data and shared it with Baidu's China server [3].

Palo Alto Networks team Unit42 used an ML-based spyware detection system to perform a root cause analysis about the data leak issue. They found that the application sent some data to a destination IP address in China. Furthermore, the data included a field called "devinfo" which was encrypted and didn't seem readable directly at first sight. Thus, on further analysis, it was found that Baidu's push SDK was fetching all the user data in terms of their International Mobile Equipment Identity (IMEI), International Mobile Subscriber Identity (IMSI), MAC address, mobile model, etc., and then combining all these data into a single field named "devinfo" which was found earlier. However, before sending all these details to their servers, they perform data encryption using Advanced Encryption Standard (AES) encryption [3].

INFORMATION COMPROMISED

Millions of users were using these apps; thus, a vast data leak had already occurred. These apps from Baidu were sharing users' data to their servers in China without seeking permission from the user to access their data. As a result, the apps were leaking the following user data [5]:

International Mobile Equipment Identity (IMEI) number
International Mobile Subscriber Identity (IMSI) number
Screen resolution
Network (Wi-Fi, 2G, 3G, 4G, 5G)
Carrier (Telecom Provider)
Android ID
Phone MAC address
Phone model

The data collected is very sensitive and can be used in various ways, including monitoring targeted people and spying. A mobile phone's IMEI number is a unique identifier for each manufactured mobile phone. It's a 15-digit code used to identify devices on a mobile network. This number remains the same even when a user changes their smartphone because it is linked with the user's telecom provider and thus is connected with the mobile's SIM card (also known as the subscriber identity module). Mobile carriers use it to track devices on their network. Authorities or authorized organizations can use the IMEI to track a phone's location, though this typically requires legal permission. Some services can also unlock a phone from a specific carrier using the IMEI number, though this should be done legally. Manufacturers and service centers can use the IMEI to check the device's warranty status and provide technical support. Some apps or services might use the IMEI to authenticate a device, ensuring it's legitimate and authorized for use.

A cyberattacker who obtains people IMEI number used for illegal activities like cloning the device, reporting that the smartphone has been stolen from the manufacturer. Consequently, there is a chance that the provider can disable the device and block its access to the network [6].

A cyberattacker can trace the user's movement [3]. The cyberattackers might also use the leaked data to intercept various phone calls or text messages. Moreover, if the encryption mode of such communication is weak, like in plain text, the cyberattacker might attempt to modify the message sent and create a man-in-the-middle attack.

Using the mobile phone's MAC address, the cyberattacker can also attempt to hack the user's smartphone. Furthermore, they can try to bypass multiple user authentications and get access to WhatsApp or other such applications with a device ID for verification [7].

HUMAN FACTORS ANALYSIS

In the digital world, where world data plays a crucial role, people need to monitor how different applications collect their data and manage or use the data collected when using the applications. The providers of the applications that launched their software or mobile-based applications need to deliver software that respects the user's privacy. The application should not unnecessarily demand information from the user that is irrelevant to that application's intended purpose. For example, a regular language translator mobile application should not ask permission to access the user's live location because it has nothing to do with the application's intended purpose: to translate text from one language into another.

Non-Reputed Companies

Many small-scale companies sometimes release mobile applications that deliberately or mistakenly collect user data, violating users' privacy. Unfortunately, these companies are not publicly recognized. As a result, people are often unaware that downloading apps published by such non-reputed companies might represent risks. Most commonly, such fraudulent applications are game-based.

Attractive but Risky Games

Some game apps have an excellent visual interface and interactive gameplay, making the users hooked on their app and repeatedly use it. Usually, people agree to all the disclaimers made by the app creator and all the permissions the app would need. The attackers use this human behavior to trick the user into agreeing to their terms of usage. Thus, they tend to give permissions to their phone book, camera, location, photos, and many more areas that have nothing to do with the intended use of that application.

Application Permission Access

Another issue is giving the application permission to collect what should not have been collected. Consequently, users need to pay more attention to details when bombarded with much information. Most users are informed to read the privacy policy the mobile application developer states they are trying to install on their smartphone. Still, it is very long, and people always rush through and agree to all its terms and conditions.

They also agree to provide all permissions in such areas, even when an app should not ask for permission to access photos if it is just a text editor app. Thus, it should create a red flag-based alert for the user when they see something like this. But people tend to overlook such important things and thus manage to give permissions to services to which the app shouldn't have access, and as a result, a data leak occurs.

In this case, users should have been prompted to grant permission to access their phone's IMEI number, IMSI number, and other sensitive information. However, the app might have requested permission for a different service, through which it automatically gained access to these sensitive mobile details indirectly.

Apps Permission User Awareness

Most apps require permission to access services like the internet, camera, etc. Still, the user should be alerted in such situations and deny permissions before using the app.

Users can install it on their smartphones. But when the user opens the app and accesses a particular feature that might require permission to access the user location, Bluetooth, etc., the permission should be granted; otherwise, the permission should be denied. Instead, permission is granted for all requests made by the app beforehand. In that case, the chances are that even when a user is not using that app, the app might run some services in the background, which might leak user data to its servers without the user being informed about it.

Users also tend to download free apps and are not worried about what permissions that app might take on the back door. The users are happy to use that app, providing excellent services for their current tasks free of cost. Keeping these things in mind, the app developers designed the app so that a maximum number of people would download it as it would be free. Still, later, the users needed to pay a hefty amount because of the data leak that might take place from that app. The two Baidu apps were free for any Android user. Thus, many people downloaded the apps. As a result, they started using those apps to get private and sensitive data of the user via those apps.

Privacy Policy and Terms of Use

Most people do not read the privacy policy and terms of use, and they do not agree with them without knowing what they have agreed to. Not reading does not mean that the users do not care about these policies or their privacy, but instead, it shows that these agreements are structured in a language and format that makes it difficult to read and understand. A significant number (over 81%) reported that they do not read because of the lengthy time it would take and because the agreements are not easy to read. Then we asked participants, "Have you ever rejected a mobile app request to access your contacts, camera, or location?" People (92%) most often have made a judgment in denying the request [8].

Consequently, we can assume that if the terms of use and privacy policy are presented clearly, underlining what users must give up in their privacy information, users can make a conscious decision about whether to use a service.

One might question why the "term of usage" or "privacy policy" is too long to read. It is written in a language that people cannot easily understand, and it is delivered in a sort of hidden UI. Of course, it is possible to present them in a simple, easy-to-understand, and straightforward UI. But should we assume that the language, length, and access are simple because software/service providers prefer users not to read them?

CONCLUSION

The Baidu case is essential, considering the number of users affected. However, this type of issue is not isolated. It requires users' awareness when using apps and many IoT devices, which these days are used in many households. The applications that run these devices commonly use the apps. The users must investigate the information the app collected and where it is sent. The user's awareness is paramount to cybersecurity in these types of cases.

REFERENCES

1. Whitney L. (2020). "Baidu Android Apps Caught Leaking Sensitive Data from Devices," TechRepublic, November 24, 2020. https://www.techrepublic.com/article/baidu-android-apps-caught-leaking-sensitive-data-from-devices/.
2. Taylor D. (2018). "Baidu SEO: A Guide to SERP Features Ranking Signals," Search Engine Journal, June 20, 2018. https://www.searchenginejournal.com/baidu-serp-features-ranking-signals-guide/257081/.
3. Achleitner S., Chengcheng Xu Ch. (2020). "Android Apps Leaking Sensitive Data Found on Google Play with 6 Million U.S. Downloads," Palo Alto Network, November 24, 2020. https://unit42.paloaltonetworks.com/android-apps-data-leakage/
4. Brewster T. (2020). "Warning: Banned Baidu Apps Exposed 'Sensitive' Data on Millions of Android Phones," Forbes, November 24, 2020. https://www.forbes.com/sites/thomasbrewster/2020/11/24/warning-banned-baidu-apps-exposed-sensitive-data-on-up-to-14-billion-android-phones/.
5. Lakshmanan R. (2020). "China's Baidu Android Apps Caught Collecting Sensitive User Data," The Hacker News, November 24, 2020. https://thehackernews.com/2020/11/baidus-android-apps-caught-collecting.html.

6. Verizon. (2021). "What to Understand about IMEI Numbers," Verizon. https://www.verizon.com/articles/what-to-know-when-buying-a-used-phone/ (accessed 2021).
7. Hackingloop. (2016). "Spoofing MAC Address on Android Mobile Phones," Hackingloop.com, March 7, 2016. https://www.hackingloops.com/spoofing-mac-address-on-android-mobile-phones/.
8. Moallem A. (2018). Do you really trust "privacy policy" or "terms of use" agreements without reading them? In D. Nicholson (Eds.), *Advances in Human Factors in Cybersecurity. AHFE 2017. Advances in Intelligent Systems and Computing*, vol. 593, pp. 290–295. Cham: Springer. https://doi.org/10.1007/978-3-319-60585-2_27.

14 Zoom

ABSTRACT

This chapter will focus on the Zoom case in which the attackers fraudulently acquired user credentials to access Zoom accounts, sold half a million Zoom account information, and analyzed it from a human factors point of view.

INTRODUCTION

Zoom Video Communications, Inc. is an American communications technology company based in San Jose, California. It provides teleconferencing, telecommuting, distance education, and social relations through a cloud-based peer-to-peer software platform. With the start of the COVID-19 pandemic, the ease of use of this software platform became very popular among many users, companies, and universities.

In early 2020, with the global COVID-19 pandemic, online meeting platforms such as Zoom increased. Because of the lockdown and the work-from-home setting, online meetings were the primary way business, professional, and educational communication took place.

In the spring of 2020, Zoom was subject to multiple cybersecurity incidents. The security issues came into the limelight on March 30, 2020, with the U.S. Federal Bureau of Investigation (FBI) announcing that they were investigating Zoom bombing attacks.

The attackers fraudulently acquired user credentials to access Zoom accounts and sold half a million Zoom accounts [1]. After the users' credentials were sold for as cheap as one cent per account in the dark web markets [2], with such account information, hackers could join unauthorized Zoom meetings of organizations such as Chase Bank, Bank of America, and Stanford, steal sensitive information, spying and do Zoom bombings, i.e., the unwanted, disruptive intrusion, generally by Internet trolls, into a video conference call [1,3].

CASE DESCRIPTION

The hackers could use the Credential Stuffing technique (automated injection of stolen username and password pairs into website login forms) to fraudulently gain access and sell half a million Zoom accounts [1]. This method uses account credentials obtained in previous breaches and used against Zoom. Knowing that many users do not change their passwords or recycle and reuse old passwords and logins, the credential stuffing attacks have successfully breached the accounts of Zoom's user base.

After the accounts, attackers could launch Zoom bombings, conduct spying, obstruct Zoom meetings of professional establishments and universities, and steal information.

DOI: 10.1201/9781003461739-19

A list of successful logins was created using credentials that leaked in previous data breaches, including the victim's email address, password, and personal meeting URL.

WHAT WE KNOW ABOUT THIS CASE

According to the press [4], Zoom misrepresented one of its features as end-to-end encryption. This type of encryption ensures that video data remains encrypted throughout the transfer, making it inaccessible even to Zoom. However, Zoom later clarified that they do not use end-to-end encryption. Instead, Zoom employs a combination of TCP and UDP connections. TCP connections are secured with TLS, while UDP connections are encrypted using AES with a key exchange over a TLS connection. The encryption Zoom uses to protect meetings is TLS, the same technology that secures HTTPS websites. This means the connection between the Zoom app on a user's PC or phone and Zoom's web server is encrypted similarly to a web browser's connection to a website. Nevertheless, transport encryption is different from end-to-end encryption, as the Zoom service can access all unencrypted video and audio content of Zoom meetings.

It was revealed that Zoom's privacy policy allowed the collection and sharing of user data, including personal information and usage patterns, through third-party services. This raised concerns about data privacy and user consent [5].

The Verge (an American technology news website) [6] also reported that Zoom grouped contacts with the same email domain into a "Company Directory" to facilitate the search for a specific person in a company with employees on Zoom, but the app has also grouped some people together who signed up for the service with a personal email, meaning an affected user might be able to see the personal email addresses and photos of people with their same domain in their Company Directory, even if none of those people are actually a colleague.

The attackers apparently created a list of successful logins using credentials leaked in previous data breaches. This list, consisting of the victim's email address, password, and personal meeting URL, was then used to sell successful logins to hackers [7].

Hackers have widely shared various configuration files for the tool and have been used in multiple stuffing attacks. The hackers also used the stress test tool to test the login of the user credentials obtained. Hackers used multiple different IP source addresses for the tools. They used the lag time between the login attempts to circumvent the Denial-of-Service (DoS) prevention mechanism that Zoom may have implemented on their server side. With a successful login attempt, zoom servers sometimes provided the hacker with meeting IDs, times, and other account information. Finally, the next step involved aggregating the successful login credentials into a new database for Zoom, which was then sold on the dark web.

From what has been reported, we can summarize the security issues into several types: problems related to the Zoom Cyberattack and issues related to the Zoom application regarding users' privacy.

Phishing Emails

Three different types of phishing emails were sent. The first type had the subject "Zoom Account," which included a message welcoming users to Zoom and then asking them to click a link to activate their account. When the user enters login credentials, the attackers steal the user's credentials and thus have access to all the user data in their account.

The second type of email was titled "Missed Zoom Meeting," which informs the user that they missed a conference call and includes a link to check the conference call.

The third type of email included subject lines such as "Meeting is canceled. Can we do a Zoom call instead?". This scam targets IT, technology, industrial, and marketing companies with malware to get access to the usernames, passwords, and credit card information of the users.

Fake Domains

According to Check Point's threat research team [8], there has been an increase in Zoom domains; 1,700 were created in early January 2020, and 425 domains have been registered in seven days. Furthermore, 70 of those domains have been identified as fake domains. They were used to carry out various cyberattacks. Since Zoom is used by 60% of Fortune 500 companies and has more than 50 million downloads, the data of many users was at stake.

Zoom Bombing Issues

Finally, a new phenomenon called Zoom bombing was used. Hackers infiltrate video meetings and often shout racial slurs, abuses, or threats. If the Zoom reference number is shared on social media and the meeting host has not set the screen-sharing setting to host-only, the hackers can take control of the host's screen and bomb it, showing disturbing material to the meeting guests. If file sharing has not been disabled, hackers can even share malware.

Zoom Users' Privacy

Zoom has a feature to track the meeting audience's attention, allowing the host of the meeting to see if a user has clicked away from zoom for more than 30 seconds. The meeting host can monitor the audience and their activities to check if someone is really attending the meeting and track meetings where they were not present. This is a severe privacy invasion issue [9].

Zoom Sells User Data to Facebook

According to a report from Motherboard, the Zoom iOS app sends analytics data to Facebook when users open it, even if they do not have a Facebook account. For example, when users open the app, their time zone, location, and device details are sent to Facebook. After a lawsuit, Zoom fixed the issue by removing the code that caused this data leak to Facebook [10,11].

Zoom Routing Calls through China

Another issue that was also discovered was issued with the encryption keys. According to the University of Toronto's Citizen Lab, during certain calls, requests for encryption keys of non-China users were routed to China and issued through servers in China. Since China does not have strict data privacy laws, they may demand Zoom to decrypt the calls, thereby revealing all the users' information at risk. For example, during the Zoom meeting test of two users, one in the United States and one in Canada, it was found that the AES-128 key for encryption and decryption was sent to one of the participants from a Zoom server located in China's capital, Beijing [12].

INFORMATION COMPROMISED

This security breach compromised about half a million user accounts. Additionally, hackers have widely shared various configuration files for the tool and used them in multiple stuffing attacks.

WHAT CYBERATTACKERS CAN DO WITH HACKED DATA?

According to Newsweek, more than 500,000 Zoom account credentials are being sold on the Dark Web for Less than a penny each [13].

Some other news report suggests an increase of unauthorized users "bombing" Zoom meetings by using them when sensitive company information and trade secrets might have been discussed. Social engineering techniques could also be used with a Zoom credential, such as setting up a fake business meeting and impersonating the Zoom used to obtain sensitive and intellectual property information from the other party. Denial of service attacks can also be used with the stolen Zoom credentials by joining a meeting, blasting music, and preventing the meeting from occurring.

Moreover, attackers used the panic and desperation of the COVID-19 situation to start phishing campaigns.

Phishing emails and messages have been sent to Zoom users and in Zoom meetings regarding the potential cure, financial relief, and information on disease symptoms. While preying on the users' fear, such phishing attacks might have been successful [14].

HUMAN FACTORS ANALYSIS

Users tend to recycle and use old logins and passwords. As a result, some Zoom users use old passwords that they may use for their email and social media. Because these credentials are old and have been obtained from prior breaches (such as Facebook, Target, and Marriott), the price is meager and sometimes can be obtained for free on the dark web. Second, hackers use various stress testing and application load testing tools. Some of these tools are legitimate tools used by testers. They configure such devices to point to Zoom servers.

If old passwords had not been used, this attack would not have happened, and this data breach could have been avoided if Zoom had implemented a decent security policy.

From a Human factors point of view, the only prevention mechanism for such attacks is to increase users' awareness about old password issues. They need to create and change their login information in the app. During the login process, it would be helpful to enforce the users to change their password and recommend using two-factor authentication, which can circumvent such stuffing attacks.

Credential stuffing is a cyberattack in which credentials obtained from a data breach on one service are used to attempt to log in to another unrelated service. Credential stuffing attacks are successful because some users use one strong password across multiple platforms. Suppose Zoom could scrape the dark web for existing user logins and against previous data breaches. In that case, Zoom could have enforced the users to change their login information either in the user interface of the login process or through secure channel to the users.

Moreover, UI reminders and recommendation messages for two-factor authentication can help convince the user to enable it. Another prevention method will be to search for the users' email or any other identification information in any recent data breaches of other companies. If such accounts have been found, Zoom can warn users that their information has been exposed in a different breach and that they should update their login information.

There are also human factors that are related to the management for not requiring encryption and implementing security features and robust encryption protocols, and their role after this cyberattack was their misleading claims about its encryption standards and not being transparent in their communication about security features.

Zoom IT personnel's role in the lack of patch management, which requires regular updates to patch security vulnerabilities.

CONCLUSION

During the COVID-19 pandemic, the surge in remote work created a significant demand for video conferencing, leading to a substantial increase in Zoom's user base [15,16]. However, the increase in Zoom usage was not parallel with more security measures for user data. Thus, Zoom was subject to various cybersecurity incidents due to Zoom system vulnerabilities, a lack of attention in handling user data, and the protection from intruders who joined users' meetings [17].

In summary, looking at this case shows that human factors that played an important role in the Zoom cybersecurity incident were:

The management role for not requiring encryption and implementing security features and robust encryption protocols and their misleading claims. Also Zoom IT personnel's role in the lack of patch management.

Users' lack of awareness of recycling and using old logins and passwords that would have protected them from these incidents [18].

REFERENCES

1. Dane Wider D. (2020). "Zoom Gets Stuffed: Here's How Hackers Got Hold of 500,000 Passwords," Forbes, April 28, 2020. https://www.forbes.com/sites/daveywinder/2020/04/28/zoom-gets-stuffed-heres-how-hackers-got-hold-of-500000-passwords/?sh=555f30d5cdc4.
2. Ikeda S. (2020). "Half a Million Zoom Accounts Compromised by Credential Stuffing Sold on Dark Web," CPO Magazine, April 27, 2020. https://www.cpomagazine.com/cyber-security/half-a-million-zoom-accounts-compromised-by-credential-stuffing-sold-on-dark-web/.
3. Singer N., Nicole Perlroth N., Krolik A. (2020). "Zoom Rushes to Improve Privacy for Consumers Flooding Its Service," The New York Times, April 8, 2020. https://www.nytimes.com/2020/04/08/business/zoom-video-privacy-security-coronavirus.html.
4. Micah Lee M., Grauer Y. (2020). "Zoom Meetings Do Not Support End-to-End Encryption Despite Misleading Marketing," Intercept, March 31, 2020. https://theintercept.com/2020/03/31/zoom-meeting-encryption/.
5. FTC. (2020). "FTC Requires Zoom to Enhance Its Security Practices as Part of Settlement," Federal Trade Commission, November 2020. https://www.ftc.gov/news-events/news/press-releases/2020/11/ftc-requires-zoom-enhance-its-security-practices-part-settlement.
6. Jay P. (2020). "Zoom Is Leaking Some User Information Because of an Issue with How the App Groups Contacts," The Verge, March 31, 2020. https://www.theverge.com/2020/3/31/21201956/zoom-leak-user-information-email-addresses-photos-contacts-directory.
7. O'Flaherty K. (2020). "Zoom's 200 Million Users Are Facing a Serious New Threat," Forbes, April 20, 2020. https://www.forbes.com/sites/kateoflahertyuk/2020/04/20/zooms-200-million-users-are-facing-a-new-threat-heres-what-to-do/#27673049b83d.
8. CheckPoint. (2020). "COVID-19 Impact: Cyber Criminals Target Zoom Domains," Checkpoint Blog, March 30, 2020. https://blog.checkpoint.com/2020/03/30/covid-19-impact-cyber-criminals-target-zoom-domains/.
9. Scroxton A. (2020). "Coronavirus. Warning over Surge in Zoom Security Incidents," Computer Weekly, 10 April, 2020. https://www.computerweekly.com/news/252480806/Coronavirus-Warning-over-surge-in-Zoom-security-incidents.
10. Cox J. (2020). "Zoom iOS App Sends Data to Facebook Even if You Don't Have a Facebook Account," Vice, March 26, 2020. https://www.vice.com/en/article/k7e599/zoom-ios-app-sends-data-to-facebook-even-if-you-dont-have-a-facebook-account.
11. Cox J. (2020). "Zoom Removes Code That Sends Data to Facebook," Vice, March 2, 2020. https://www.vice.com/en/article/z3b745/zoom-removes-code-that-sends-data-to-facebook.
12. Wood Ch. (2020). "Zoom Admits Calls Got 'Mistakenly' Routed through China," Insider, April 6, 2020. https://www.businessinsider.com/china-zoom-data-2020-4.
13. Czachor E. (2020). "More Than 500,000 Zoom Account Credentials Being Sold on Dark Web for Less Than a Penny Each," April 13, 2020. https://www.newsweek.com/more-500000-zoom-account-credentials-being-sold-dark-web-less-penny-each-1497624.
14. Hodge R. (2020). "Zoom Security Issues: Zoom Buys Security Company, Aims for End-to-End Encryption," CNET, May 8, 2020. https://www.cnet.com/news/privacy/zoom-security-issues-zoom-buys-security-company-aims-for-end-to-end-encryption/.
15. Carman A. (2020). "Why Zoom Became So Popular," The Verge, April 3, 2020. https://www.theverge.com/2020/4/3/21207053/zoom-video-conferencing-security-privacy-risk-popularity.

16. Paul K. (2020). "Zoom Is Malware: Why Experts Worry about the Video Conferencing Platform," The Guardian, April 2, 2020. https://www.theguardian.com/technology/2020/apr/02/zoom-technology-security-coronavirus-video-conferencing.
17. Cox J. (2020). "Zoom Removes Code That Sends Data to Facebook," Vice, March 27, 2020. https://www.vice.com/en/article/z3b745/zoom-removes-code-that-sends-data-to-facebook.
18. Osborne Ch. (2020). "Zoom Security: Your Meetings Will Be Safe and Secure if You Do These 10 Things," SDNET, April 22, 2020. https://www.zdnet.com/article/make-sure-your-zoom-meetings-are-safe-by-doing-these-10-things/.

15 LinkedIn

Abstract

This chapter focuses on the LinkedIn data breach that affected a significant portion of LinkedIn's user base, making it one important data breaches. It analyzes it from a human factors point of view.

INTRODUCTION

LinkedIn is an American service business and employment-oriented online service that operates via websites and mobile apps. It has over 740 million registered users worldwide. As more and more people use social networking websites and applications to stay connected with their family, friends, and coworkers, millions of people use social media platforms like LinkedIn to search for jobs, update their work profiles, build professional portfolios, and network.

LinkedIn has experienced multiple cybersecurity incidents over the years. In 2012, LinkedIn experienced a significant data breach where approximately 6.5 million user passwords were stolen and posted online. In 2016, it was reported that a hacker was selling data of around 117 million LinkedIn users on a dark web marketplace. The data included email addresses and hashed passwords obtained through previous breaches. LinkedIn invalidated affected passwords and encouraged users to reset them [1,2].

In April 2021, LinkedIn experienced a significant data breach that compromised the personal information of millions of its users. The breach was a severe threat to cybersecurity, and it's estimated that it impacted more than 500 million users or roughly 40% of LinkedIn's entire user base [3]. This data breach affected a significant portion of LinkedIn's user base, making it one of the largest data breaches in history [4]. The leaked information can be used for various malicious purposes, such as phishing attacks, identity theft, and even cyberstalking.

CASE DESCRIPTION

LinkedIn provides its users services like connecting and interacting with other users to build their resumes and improve their network. As stated on techcrunch.com [2], this case was revealed by Ireland's Data Protection Commissioner (DPC) in a report that covered all activities in the six months of 2018 (January to June). The investigation list included companies like Facebook, Yahoo, and WhatsApp, where these incidents were reported. However, LinkedIn was not part of that list until 2017, when one user complained about LinkedIn's practices of targeting non-users on social media. LinkedIn case is not really a cyberattack case but more a privacy breach that is the unauthorized or improper access to, use, creation, collection, disclosure, retention, or disposal of personal information.

DOI: 10.1201/9781003461739-20

LinkedIn used non-users' email addresses without their consent to attract more users. They used these email addresses in hashed form and sent their targeted user advertisements on Facebook. After the investigation, the DPC's commissioner said, "The complaint was ultimately amicably resolved, with LinkedIn implementing some immediate actions to cease the processing of user data for the purposes that gave rise to the complaint." The DPC also conducted further audits on LinkedIn to check for other data concerns similar to this [5].

Following up on this complaint and investigation, as mentioned on techcrunch.com [2], Denis Kelleher, who is head of privacy in LinkedIn, said, "We appreciate the DPC's 2017 investigation of a complaint about an advertising campaign and fully cooperated," and "Unfortunately, the strong processes and procedures we have in place were not followed and for that we are sorry. We have taken appropriate action and have improved how we work to ensure that this will not happen again. During the audit, we also identified another area where we could improve data privacy for non-members, and we have voluntarily changed our practices." LinkedIn accepted its involvement in using email addresses without the user's consent. However, LinkedIn was not fined because, at that time, GDPR regulators didn't have the power to enforce fines for such activities [6].

WHAT WE KNOW ABOUT THIS CASE

LinkedIn's primary motive for violating data protection appears to be profit-driven, aiming to increase the number of users who engage with their platform by utilizing their specific information. These issues were highlighted when a non-LinkedIn user complained that LinkedIn was using his email ID to target him with Facebook advertisements. After this, many other people complained about LinkedIn for the non-consensual use of their email addresses.

The Irish DPC investigated this and found some issues with how LinkedIn processes users' personal data. Also, the source body from which LinkedIn accessed these 18 million email addresses has yet to be discovered. As described in the CPO Magazine [1], the DPC's report said that U.S. LinkedIn had used 18 million users' data to target them on Facebook, in the absence of LinkedIn Ireland. After completing this investigation, DPC also performed a further audit on LinkedIn, where they learned about LinkedIn's use of social graph-building algorithms to build their network. They were making these networks to help users avoid the painful process of building networks from scratch. As a result of the findings, the Irish data controller asked LinkedIn to stop the computing process and delete all records of non-consensual data they had.

INFORMATION COMPROMISED

The important known information compromised in this case was the email addresses of 18 million non-LinkedIn users. Though the source of this leak is unknown, it can be from any other social networking website, such as Facebook. Additionally, per the Malwarebytes [7] blog on why you should not use LinkedIn, LinkedIn may monitor its users' actions. As a result, it can use the information shared by other people.

It can also be the case that Facebook itself sold the information of its targeted users to LinkedIn in return for money or other business favors. If this is the case, email addresses and other private data like mobile numbers, photos, messages, etc., may also have been leaked.

WHAT CYBERATTACKERS CAN DO WITH HACKED DATA?

If hackers get a hold of the email addresses of people, then they can exploit it in many ways, as follows [3]:

Scamming Contact List

Once hackers get the email address, they also gain access to the digital contact list. Then hackers can send spam and phishing emails to people disguised as users' friends that, if clicked, the computer is infected, or the email can also ask for money, which will go to the hacker.

Mass Email Scams

After the hackers are done using the contact list, they can start sending large amounts of emails from users' accounts to unknown recipients. This phishing attack can infect other computers or ask for a ransom in return for access. This action will be traced back to the user to whom the phishing emails were sent. In this way, the hackers stay safe.

Infiltrating Devices and Programs

Once the hacker can access users' email, attackers can infiltrate users' computers even if it has enabled the Health Insurance Portability and Accountability Act (HIPAA) encryption. They can access data on the computer, connected mobile devices, cloud, and shopping account details. Many people store passwords of all accounts in one place on a computer. If they are hacked, the hacker can reset all email passwords. Hackers can also get access to credit cards and social media accounts.

HUMAN FACTORS ANALYSIS

The case concerns how the company misused the 18 million users' emails. LinkedIn might have gotten this information from other social networking websites like Facebook, Instagram, etc.

If these emails get into the wrong hands, a crisis can be caused. Electronic mail is the most common way through which fraud occurs. Phishing emails containing links are sent to the targets from their trustworthy source but disguised. Upon clicking the link, the target can get hacked or infected.

Knowing how to spot phishing emails can stop these attacks, but only if observed. Also, we should avoid clicking unknown/unwanted links or installing malicious software. One should be more careful while using social media and always check their privacy settings before sharing private information.

CONCLUSION

In summary, human factors again were one of the main issues, including the management role in misusing users' email and phishing attacks, which resulted in installing malicious software by tricking the users into clicking phishing mail, thinking that it was from a trusted source.

REFERENCES

1. Perez S. (2016). "117 Million LinkedIn Emails and Passwords from a 2012 Hack Just Got Posted Online," TechCrunch, May 18, 2016. https://techcrunch.com/2016/05/18/117-million-linkedin-emails-and-passwords-from-a-2012-hack-just-got-posted-online/.
2. Lunden I. (2019). "LinkedIn Violated Data Protection by Using 18M Email Addresses of Non-members to Buy Targeted Ads on Facebook," Tech Crunch, November 24, 2018. https://techcrunch.com/2018/11/24/linkedin-ireland-data-protection.
3. Canales K. (2021). "LinkedIn Data of 500 Million Users Being Sold Online, Leaked Details Include Email Addresses and Phone Numbers," Business Insider, April 8, 2021. https://www.businessinsider.com/linkedin-data-scraped-500-million-users-for-sale-online-2021-4.
4. Tidy J. (2020). "How Your Personal Data is Being Scraped from Social Media," BBC News, July 2021. https://www.bbc.com/news/business-57841239.
5. Ingrid Lunde I. (2018). "LinkedIn Violated Data Protection by Using 18M Email Addresses of Non-members to Buy Targeted Ads on Facebook," TechChrunch, November 24, 2018. https://techcrunch.com/2018/11/24/linkedin-ireland-data-protection/.
6. Mayer S. (2018). "LinkedIn Crossed the Privacy Line with Targeted Facebook Advertising," CPO, November 29, 2018. https://www.cpomagazine.com/data-privacy/linkedin-crossed-the-privacy-line-with-targeted-facebook-advertising/.
7. Tsing W. (2018). "Maybe You Shouldn't Use LinkedIn," Malwarebytes Blog, April 5, 2018. https://blog.malwarebytes.com/security-world/2018/04/maybe-you-shouldnt-use-Linkedin/.

16 Spotify

Abstract

This chapter focuses on the Spotify case, which compromised information contained the users' email IDs, passwords in plaintext, and any other data validated against the users' Spotify accounts. It analyzes it from a human factors point of view.

INTRODUCTION

Spotify is a Swedish audio streaming and media service provider. On November 23, 2020, vpnMentor (providing tools for navigating VPNs and web privacy) published a report revealing that about 300,000 Spotify users might have their personal information exposed [1]. Spotify servers were not themselves compromised; the users' information was compromised by a technique known as Credential Stuffing. Credential stuffing is the automated injection of stolen username and password pairs ("credentials") into website login forms to fraudulently gain access to user accounts.

The compromised information contained the users' email IDs, passwords in plaintext, and any other data validated against the users' Spotify accounts (for example, the user's country of residence). To protect their users' accounts, Spotify implemented a Rolling Reset of the passwords of the compromised accounts.

CASE DESCRIPTION

According to the vpnMentor research lab, a pro bono service strives to help the online community defend itself against cyber threats while educating organizations on protecting their users' data [1].

On November 23, 2019, vpnMentor published a report stating that they had discovered a 72 GB Elasticsearch database that, among other things, contained login ID, password, and other Personally Identifiable Information (PII) of 300,00–350,000 active Spotify users. vpnMentor brought this to Spotify's attention and determined that Spotify's servers had not been compromised. Instead, the users' Spotify information was acquired through a process known as Credential Stuffing, a popular hacking method when a user reuses their credentials across different services [1].

Working with Spotify, vpnMentor found that the database they had discovered likely belonged to a third party that might have used it to defraud Spotify and its users. Spotify and vpnMentor determined that the hackers had used credentials from some other service. They concluded that the uncovered database was likely a subset of the larger database and only contained the entries that were valid "hits" on Spotify, i.e., had shared credentials between the compromised service and Spotify [1].

DOI: 10.1201/9781003461739-21

Spotify implemented a Rolling Reset of the compromised passwords in response to the exposed credentials. Additionally, they reset the passwords of the compromised accounts to prevent malicious login and usage of the compromised accounts [1].

WHAT WE KNOW ABOUT THIS CASE

A research team at vpnMentor discovered an Elasticsearch (a search engine based on the Lucene library) database with over 380,000 million entries, containing credentials of over 300,000 active Spotify users. Spotify and vpnMentor classified it as a possible credential-stuffing operation, but could not track the origin or intent of the hackers [1].

Credential Stuffing is a common technique to hack accounts. According to the OWASP Foundation, credential stuffing is the automated injection of breached username/password pairs to gain fraudulent access to user accounts [2]. In essence, credential stuffing allows users to use the same (or similar) ID and password to log into different services. If one service is compromised and the credentials are extracted, hackers can try these credentials against other services to see if the credentials are shared.

In this case, some unknown service was compromised, giving the hackers a compromised login credential database. The hackers attempted to log in to Spotify using the compromised credentials. They generated a database of the credentials that logged in successfully and any personal information about the users they could find.

The researchers at vpnMentor discovered this filtered database; the original database and the compromised service are still unknown [1]. While Spotify's server was never compromised, there are sure security holes in Spotify's flow that increase the impact of low-effort techniques like Credential Stuffing. An enormous hole is the lack of two-factor authentication (2FA). 2FA asks the user to provide another form of authentication, for example, a randomly generated PIN or a hardware key, to prove identity. 2FA does not prevent Credential Stuffing; it does prevent hackers from accessing any profile information attached exclusively to Spotify (for example, location). As of the writing of this report, Spotify has added 2FA as a feature "Under Consideration" [3].

INFORMATION COMPROMISED

vpnMentor did not release the complete list of information present in the database but stated the following list as compromised information: Account username and plaintext passwords for active Spotify accounts, email addresses and countries of residence.

However, Spotify's lack of 2FA means that all information attached to the compromised account can be extracted, including linked social media accounts, information about devices, and connected accounts through the Spotify family.

According to vpnMentor, several IP addresses present in the database seem to be network operators rather than the users. However, these IPs are publicly accessible and thus are not considered compromised [1]. However, the hackers still have the users' email and other personal information and could re-launch a credential-stuffing attack against some other service.

WHAT CYBERATTACKERS CAN DO WITH HACKED DATA?

Spotify's lack of 2FA meant the hackers could access all information attached to the compromised users' Spotify accounts. The impact of compromised information can range from trivial account abuse to more directed identity theft. There are also several options that attackers can exploit from users' profiles, including the following:

Financial Fraud or Identity Theft

If the users connect their social media accounts to Spotify, the hackers can track the users' identities across multiple social networks. Using the information acquired also enables the attackers to build a complex profile for each user. Then, track down and attempt to crack some social media accounts using Credential Stuffing (or other similar techniques). The hackers can create highly targeted attacks for financial fraud or identity theft depending on the completeness of the information they can gather.

Phishing or Malware

Since the hackers have access to active Spotify account emails, they can create targeted phishing attacks to gather more sensitive information, such as credit card or bank details. If successful, this can snowball into full identity theft or financial fraud.

Along similar lines, hackers can imitate Spotify and trick users into clicking malicious links in the email, which inject malware into the users' computers.

Account Abuse

The compromised active Spotify accounts can be used to subscribe to Spotify's premium service, enabling hackers to sell the accounts in an open market where unauthorized people can use the premium service.

HUMAN FACTORS ANALYSIS

The case illustrates how humans can be the weakest link. While Spotify is not a security model, its servers remained uncompromised throughout the case. The surface of the attack was entirely orthogonal to Spotify.

The root of vulnerability lies in the users themselves. Specifically, users who use the same credentials across multiple services. At the time when every service requires credentials to log in, these credentials invariably being an email ID and a password, it is not uncommon for users to default to using the same credentials across services. This reduces the users' cognitive load as they have fewer combinations of IDs and passwords to remember, making them vulnerable to credential-stuffing attacks. This is one of the primary reasons security experts recommend using a password manager and creating strong passwords that a brute-force attack cannot easily guess.

Additionally, people's tendency to reuse credentials is not uncommon, making them vulnerable due to the Credential Stuffing of compromised accounts. The ordinary nature of the attack means that Spotify needs to take steps to prevent commonplace attacks like credential stuffing. We discussed how a simple 2FA could have prevented the hackers from collecting the user's personal information.

CONCLUSION

In conclusion, this case's main human factors issues were related to a need for more awareness of users when using the same credentials across multiple services. Also, a lack of understanding of the importance of using 2FA authentication that can provide them with an extra level of securing their accounts.

REFERENCES

1. Fawkes G. (2020). "Report: Spotify Targeted in Potential Fraud Scheme," vpnMentor, November 23, 2020. https://www.vpnmentor.com/blog/report-spotify-scam/.
2. Mueller N. (2021). "Credential Stuffing," OWASP, 2021. https://owasp.org/www-community/attacks/Credential_stuffing.
3. BrainStation (2024). "Two Factor Auth (2FA)," BrainStation, Accessed December, 15, 2024. https://brainstation.io/cybersecurity/two-factor-auth.

Part Six

Trust

- Uber Technologies
- Equifax
- Office of Personnel Management Data

17 Uber Technologies

ABSTRACT

This chapter is focused on the Uber Technologies case. After describing the issue and what we know about it, we will analyze it from a human factors point of view.

INTRODUCTION

Uber is an American multinational ride-hailing company offering peer-to-peer ride-sharing, ride-service hailing, food delivery, and a micro-mobility system with electric bikes and scooters. The ride-sharing market is the growing personalized transport. Uber holds a significant share of this market. In addition, Uber owns a ride-hailing platform that allows users to call an on-demand cab and connect the user to the cab driver [1]. In its early days, Uber grew exponentially and became the most extensive ride-hailing service in the United States.

The breach happened in October 2016 when two attackers hacked data for about 57 million users. The breach also granted access to the personal data of about 4.1 million drivers [2]. The hacked user data included names, email addresses, phone numbers, driver's licenses, and how drivers drive. This attack was a significant data breach at Uber, which compromised the personal information of millions of Uber riders and drivers. This case led to widespread criticism and a significant penalty from the Federal Trade Commission [3].

Uber collects personal information from users, such as names, email addresses, mobile numbers, and credit card information. Also, the application requires access to the user's and the driver's location to call a cab at the user's current location [1]. The drivers are essentially Uber employees, so they also have all the drivers' employment-related information, including driver's licenses, vehicle registration, and social security number (SSN), which is highly confidential.

All users' and drivers' information are usually stored in a secure environment on the cloud. However, since the cloud is theoretically accessible from anywhere in the world if you have the necessary security clearance, it poses unique security challenges.

Although the breach happened in 2016, it was never revealed until 2017. Uber paid the hackers $100,000 as a "bug bounty" reward to destroy the data they obtained to avoid notifying the users or the regulators. Later, Uber paid the Federal Trade Commission $148 million for this case [4].

This chapter is focused on this case. After describing the issue and what we know about it, we will analyze it from a human factors point of view.

DOI: 10.1201/9781003461739-23

CASE DESCRIPTION

In November 2016, two attackers who were not employees gained access to the company database [5]. The attack was a significant security breach as it compromised the personal information of 57 million riders and Uber drivers.

Uber later said, "We have seen no evidence of fraud or misuse tied to the incident" to its riders [6]. In addition, the company was monitoring the accounts affected by the breach. However, reports later revealed that Uber did not promptly disclose the violation. An investigation led by the FTC (Federal Trade Commission) also found that Uber failed to comply with personal data security practices. This case prompted the FTC to impose a heavy $148 million penalty on Uber.

Uber later also confirmed that it paid the hackers a ransom of $100,000 to conceal the breach and delete the downloaded data. The company's failure to notify the related regulators and users prompted hard action against Uber's security team [7].

WHAT WE KNOW ABOUT THIS CASE

Uber's database was stored in the Amazon web server cloud service. Unfortunately, in October 2016, two attackers gained access to the database by cracking the database password through brute force [5]. Since the database stores users' data unencrypted, hackers can download sensitive data like names and email addresses in plain text.

It was also later found out that the web server used to store the data had a public IP accessible from any IP address. Thus, it was a significant lapse from the security team of Uber as data was kept unencrypted. Cybersecurity firm BullGuard's CEO said this mistake was "unforgivable" [2].

In late 2017, when Uber's CEO announced the data breach, more research revealed that it was because of the careless behavior of an employee. The attack started when the hackers were able to get access to the employee's GitHub account. GitHub is a cloud service where software engineers can usually collaborate and work on the same code for a single project. The hackers accessed the company's code repository once the GitHub account was compromised due to the employee's negligence. The hackers found a lot of data in the codebase and managed to get access to credentials for Amazon Web Servers, which belonged to Uber.

Later, it was reported that two Uber developers had stored the credentials for the company's data stores in their code GitHub.

The major problem with this data breach was that Uber did not notify the regulators or users, although the attacker destroyed the data. However, this case has affected Uber's credibility and got much media attention. Furthermore, they did not report this data breach by exposing their users to potential financial fraud or identity theft. This revelation led to several lawsuits being filed against the company. Later, the company was ordered to pay $148 million in compensation for the data breach. Although the company had "obtained assurances" that the data was destroyed, if it was not, then it could have led to significant problems for the users and the drivers.

INFORMATION COMPROMISED

The attackers obtained the names, email addresses, and phone numbers of 57 million Uber users. This hack also obtained the driving license numbers of about 4.1 million drivers.

Uber assured us that the breached data did not include trip location history, credit card numbers, bank account numbers, Social Security numbers, or dates of birth.

For riders, personal information like names and email addresses was compromised. Payment information, such as credit card numbers, was also compromised for some riders, which is highly sensitive [4]. The ride history of some riders was also compromised by revealing their home address for those riders.

The personal information of Uber drivers was also compromised, including their vehicle information, such as driver's license and vehicle registration [3]. In addition, the repair or accident history of the drivers, in some cases, was also hacked.

The information acquired by attackers can be used for identity theft or financial fraud since the driver's license numbers of drivers were stolen. In addition, the attackers can also use stolen data for phishing attacks.

The attackers obtained information like bank details, credit card numbers, and other sensitive data, exposing drivers and users to fraudulent transactions. Using this data, they can also obtain personal information and health information.

WHAT CYBERATTACKERS CAN DO WITH HACKED DATA

In this case, the extent of the leaked information is so huge that hackers can do anything they want. However, they can primarily blackmail the company, which they did, and ask for a ransom for not revealing the data. Uber paid the ransom to hackers to conceal the breach in this case.

Second, they can use personal information for phishing attacks against users. For example, they can trick the users into giving out credit card information or account credentials. Hackers can also use the users' ride history to find their home addresses and other sensitive information. Hackers can also sell personal information, such as users' mobile numbers, to third-party companies for advertisements.

HUMAN FACTORS ANALYSIS

Several areas are related to the human agents' behavior and vulnerability based on what we know.

Management Behavior

Attacks on large-scale organizations like the Uber case might indicate a lack of significant resources to secure the network and people's data.

Waiting to Announce: The data breach happened in 2016, and Uber officials waited till late 2017 to announce this breach. If the hackers had not destroyed the data, they could have used it for fraud.

Trusting in the Kindness of Hackers: The hackers were paid $100,000 to destroy the data and made them sign a nondisclosure agreement. Experts believe that trusting in the kindness of the hackers is not good business sense.

IT and Developer Professional Behavior

- Storing Sensitive Data in Repositories: Software developers at Uber had stored sensitive login data on GitHub. This should be avoided because if this codebase is hacked, it leads to many other things.
- Making the Same Mistake Twice: Uber developers have made the same mistake twice. In 2014, hackers found a login key left in a publicly posted codebase on GitHub, which led to the theft of 50,000 Uber drivers' data.
- **After studying** this case, the following measures can be taken to improve the security of human interaction with the Uber network:
  a. Keep the GPS location of the rider password-protected and encrypted over the network.
  b. Only specific employees can access the geolocation information using access control and authorizations.
  c. Annual employee training should be conducted for the network's security personnel.
  d. Multi-factor authentication should be used to access any rider-specific information.
  e. Regular audits should be done by a third party that would evaluate the network's security level.

CONCLUSION

Uber's data breach can be represented as a classical security breach case that involved hackers gaining access to valuable information. As a result, the company misused consumer trust and paid a hefty $148 million in penalties [8]. Also, the data breaches should be reported within the time limits provided by government agencies. Not revealing the data breach can lead to further scrutiny and loss of public trust in the company. In addition, two-factor authentication and authorization are necessary for access to sensitive company information.

REFERENCES

1. Golden Data Law. (2019). "The Uber Breach Story: On How Security Woes Can Lead to a Criminal Complaint," Golden Data Law, May 13, 2019. https://medium.com/golden-data/case-study-uber-technologies-inc-data-breach-7261484d6471.
2. Wong J. C. (2017). "Uber Concealed Massive Hack That Exposed Data of 57m Users and Drivers," The Guardian, 22, 2017. https://www.theguardian.com/technology/2017/nov/21/uber-data-hack-cyber-attack.
3. Chappell B. (2018). "Uber Pays $148 Million over Yearlong Cover-up of Data Breach," NPR, September 27, 2018. https://www.npr.org/2018/09/27/652119109/uber-pays-148-million-over-year-long-cover-up-of-data-breach

4. Hay L. (2021). "A Former Uber Exec's Indictment Is a Warning Shot," Wired, August 21, 2020. https://www.wired.com/story/uber-exec-joe-sullivan-data-breach-indictment/.
5. Whittaker Z. (2018). "Two Hackers behind 2016 Uber Data Breach Have Been Indicted for Another Hack," TechCrunch, October 25, 2018. https://techcrunch.com/2018/10/25/uber-hackers-indicted-lynda-breach/.
6. Collins K. (2017). "Uber Admits 2.7m Brits Hit by Data Breach," CNET, November 29, 2017. https://www.cnet.com/tech/services-and-software/uber-admits-2-7m-brits-hit-by-data-breach/.
7. Fair L. (2018). "FTC Addresses Uber's Undisclosed Data Breach in New Proposed Order, FDC, April 12, 2018. https://www.ftc.gov/news-events/blogs/business-blog/2018/04/ftc-addresses-ubers-undisclosed-data-breach-new-proposed.
8. Chappell B. (2018). "Twitter Uber Pays $148 Million over Yearlong Cover-up of Data Breach," NPR, September 27, 2018. https://www.npr.org/2018/09/27/652119109/uber-pays-148-million-over-year-long-cover-up-of-data-breach.

18 Equifax

Abstract

This chapter discusses the Equifax breach, from how it occurred to what is at stake. It also explores the risk this breach carries and how human–computer interaction elements have affected this case, both from Equifax's and Equifax's customers' perspectives.

INTRODUCTION

Equifax is a consumer credit reporting agency based in the United States. It is among the top three most prominent agencies in this field. It operates on a massive scale: It has 800 million personal and 88 million business consumers, generating USD 5.1 billion in annual revenue. Equifax provides demographic and credit-related data to its business-based customers. Additionally, Equifax offers credit monitoring and fraud prevention services for individual customers. The size of Equifax's operations directly impacts the scale of the data breach in Equifax (equifax.com).

In September 2017, Equifax released an official statement saying that the company was a victim of a significant cyberattack, which led to a massive data breach [1,2]. Considering the size of Equifax's operation, its data breach was considered one of the biggest in history.

Equifax withheld the knowledge of the breach for a month, leading to accusations of insider trading later, as some of the top executives sold their shares [3]. Although there have been data breaches of size and on a bigger scale than this one, the severity of the Equifax data breach was higher, considering the sensitivity of the information in question. It had personal identification tied to financial information, mostly of U.S.-based customers. It affects at least half of the U.S. population, putting them at risk of identity theft [4]. As the severity became clear, Equifax was initially criticized for the breach itself; later, it was criticized for the lackluster response. Some top officials of the company were accused of corruption as well. A software engineer was charged with confidential trading information on the customers impacted by the breach [3]. It is unclear what the hackers' motive was, as this information has yet to be directly proven to be misused anywhere. The identity of these hackers is still unknown.

This chapter discusses this breach, from how it occurred to what is at stake. It also explores the risk this breach carries and how human–computer interaction elements have affected this case, both from Equifax's and Equifax's customers' perspectives.

CASE DESCRIPTION

From May 2017 to July 2017, anonymous cybercriminal hackers accessed one of Equifax's servers. As a result, 143 million Equifax clients were impacted [5,6]. In addition, Equifax spent 1.4 billion on upgrading security after this attack [2].

 DOI: 10.1201/9781003461739-24

Three weeks after the breach became public, the CEO stepped down. In addition, the Securities and Exchange Commission charged a former Equifax executive with insider trading after he made millions selling shares before the public knew about the attack. Equifax also hired a new chief security officer.

Equifax implemented a new management system to handle vulnerability updates and verify the patch implementation following this attack [7].

After the announcement, Infosec experts started monitoring the dark web, expecting vast amounts of stolen information. However, this explanation was proved wrong and led to different ideas about the theft. In February 2020, the U.S. Department of Justice officially charged four men from the Chinese military for the attack [4]. Other hacks in the Anthem and Starwood hotels' databases were also related to the same group [8].

WHAT WE KNOW ABOUT THIS CASE

The cyberattackers initiated the first penetration from the consumer complaint web portal. Then, the attackers moved to servers other than the consumer complaint web portal. Equifax did not have segmented servers. Thus, the attackers got access to usernames and passwords in plain text, allowing them to access the information in these accounts. It seems the attack went undetected for a long time because the attackers downloaded the information in an encrypted format. However, the company still needed to renew the encryption certificate [9]. Thus, the cyberattack damage was already done when the encryption certificate was renewed, and the penetration was discovered.

The U.S. Government Accountability Office released a report diagnosing the breach. The hackers started looking for vulnerabilities in the system on March 10, 2017 [10]. Two months of research led them to Equifax's dispute portal, the platform for people to file and defend their claims. The website used the Apache Struts framework (an open-source development framework for creating Java applications). This framework has been discovered to be vulnerable to malicious codes being sent as part of the content-type header of the HTTP requests [2]. HTTP response header injection is a security vulnerability that occurs when an attacker can inject malicious content into the response headers returned by a web server. It can also lead to various attacks, such as cross-site scripting (XSS), session hijacking, cache poisoning, and phishing. Struts would then execute the malicious code, which paved the way for further intrusions. They used this vulnerability and accessed the credentials for three servers [10]. The same credentials also allowed them to access another 48 servers that contained users' personal information [2]. According to the report, it was already 76 days before the breach was even detected. The hackers slowly stole data from all 51 servers to avoid raising red flags. In addition, they pulled the data from the network in encrypted form. The breach remained undetected as the company had failed to renew an encryption certificate on one of its internal security tools [2]. Also an expired certificate should have raised an alarm which should cause an investigation that could detect a breach. Equifax finally caught up and closed the attack on June 30, 2017.

INFORMATION COMPROMISED

The Equifax statement declared that a minimum of 209,000 customers' credit card credentials were taken. Later, in March 2018, they updated that the figure to add 2.4 million U.S.-based customers. The data consisted of full names, social security numbers, date of birth, address, driving license numbers, etc.

The attack compromised the personal information of 143 million Americans, reported to have been stolen, including:

- Names, addresses, and birthdates.
- Social Security numbers and driver's license numbers.
- Credit card numbers for 209,000 customers.
- Documents with personal information used in disputes for 182,000 people were also taken.

The information was stolen using the company's claims page. The claim page was one of the 15 vulnerable fronts available [11,12].

WHAT CYBERATTACKERS CAN DO WITH HACKED DATA

After the information about the Equifax data breach was made public, the cybersecurity experts, as per protocol, actively observed the dark web for any information from this data breach to surface. According to the Federal Bureau of Investigation (FBI) of the United States, there is evidence that a country backed these attacks. The investigators also closely relate the Equifax data breach to two other breaches (the U.S. Office of Personnel Management and Marriot's Starwood Hotel). However, the data from these breaches has yet to surface. The attacks aimed to build a vast data pool on Americans. This pool could be exploited using big data techniques to uncover information on critical U.S. government officials [3].

The data acquired in the Equifax data breach is extremely valuable to attackers as they can use this information to execute a variety of fraudulent acts, including but not limited to:

- Identity theft to start new credit accounts, apply for loans, and gain other financial services in the victims' identities, which can result in severe financial loss for the victims, affecting their credit scores and a lengthy and expensive process to restore their identity.
- Synthetic identity theft involves building new identities by combining actual and fraudulent personal information.
- Account takeover to gain control of existing accounts, including but not limited to credit card accounts, investment accounts, and bank accounts.
- Conduct phishing attacks by creating convincing phishing emails or other forms of fraud using stolen personal information.

- Medical identity theft and perpetrating of medical identity and access to medical information for targeted attacks.
- Employment fraud and use of victims' personal data to apply for jobs or government benefits on behalf of the victims. As a result of misleading information in their records, victims may be denied benefits or job chances.

Ransomware threats to acquire access to corporate networks, attackers can exploit personal information to generate convincing phishing emails or other social engineering attacks.

HUMAN FACTORS ANALYSIS

From a human factors point of view, the Equifax case exploited many human errors. According to Bloomberg, the Equifax case showed typical flaws leading to a cybersecurity disaster [11]. The employees of Equifax used to speculate jokingly that their company was only a single hack from going bankrupt [5].

EQUIFAX MANAGEMENT BEHAVIOR

Equifax withheld knowledge of the breach for a month, leading to later accusations of insider trading as some top executives sold their shares.

Before this attack, Equifax was not wholly invulnerable to cyberattacks. However, while the previous attacks could have allowed Equifax to reanalyze its security approach, they failed.

It is also not unreasonable to conclude that the breach occurred due to human negligence. Two months before the attack, Equifax was informed regarding the vulnerability CVE-2017-5638 found in the Apache Struts framework [3]. In response, the Apache Software Foundation released a patch on March 7. Since Equifax was one of the many companies whose websites used this as the underlying network, the administration was promptly informed to apply the patch on March 9. On March 15, their IT department ran a series of tests to determine which systems needed the patch. However, the scans proved incompetent since they did not flag any systems. Simultaneously, they had also employed Madinat, a security consulting firm, to evaluate their systems. They even warned Equifax regarding the multiple unpatched and misconfigured systems.

CERTIFICATE RENEWAL

Another oversight on Equifax's part was regarding the renewal of certificates. Equifax had tools to decrypt, analyze, and re-encrypt internal network traffic. This process was in place primarily to sniff out any breaches like this. However, the re-encryption process depended on the public-key certificate, which expired over ten months before the attack [3]. Therefore, the network traffic was left uninspected.

The emergence of issues had been the norm at Equifax; worse, the dull and slow response was also the norm. For example, according to an ex-employee (in an unrelated incident to this breach), no action was taken when he told the company about some servers having open file-sharing ports. As a result, several months later, these servers were attacked by the Conficker worm [12]. This shows a general lack of seriousness regarding Equifax's handling of cybersecurity.

CONCLUSION

The Equifax data breach exposed the confidential information of nearly half the American population. The absence of the breached data on the dark web suggests that this attack aimed to gather information on classified U.S. government officials. Before this incident, Equifax was not completely vulnerable to cyberattacks. However, previous attacks, which could have prompted Equifax to reevaluate its security approach, were ineffective.

This lax attitude permeated the company, resulting in a culture of relaxed tolerance toward cybersecurity from a human–computer interaction perspective. The response from Equifax customers affected by this breach was also mild, highlighting a broader issue of public misunderstanding regarding data misuse. In the future, the intensity of such attacks will only increase unless Equifax addresses the serious flaws in its culture. Although the company invested $1.4 billion in improving its security post-attack by upgrading tools, software, etc., there still needs to be a program focused on creating a new policy on cybersecurity tolerance from a human–computer interaction perspective. Without addressing this, Equifax remains in grave danger of future breaches.

Considering all factors, it is not unreasonable to conclude that human negligence contributed significantly to the breach, which could have been avoided with proper protocol [3].

REFERENCES

1. Popken B. (2017). "Equifax CEO Richard Smith Steps Down after Epic Breach," NBC News, September 26, 2017. https://www.nbcnews.com/business/consumer/equifax-ceo-richard-smith-retires-after-epic-breach-n804771.
2. Farlinger J. (2019). "Equifax Data Breach FAQ: What Happened, Who Was Affected, Was the Impact?" CSO US, February 12, 2020. https://www.csoonline.com/article/3444488/equifax-data-breach-faq-what-happened-who-was-affected-what-was-the-impact.html
3. Securities and Exchange Commission. (2018). "Former Equifax Manager Charged with Insider Trading," SEC, 2018–115. https://www.sec.gov/news/press-release/2018-115.
4. Zou Y., Mhaidli A. H., McCall A., Schaub F. (2018). I've got nothing to lose. *The Proceedings of the Fourteenth Symposium on Usable Privacy and Security,* USENIX, August 12–14, 2018, Baltimore, MD. https://www.usenix.org/conference/soups2018/presentation/zou.
5. Siegel Bernard T., Hsu T., Perlroth N., Lieber R. (2017). "Equifax Says Cyberattack May Have Affected 143 Million in the U.S.," New York Times, September 7, 2017. https://www.nytimes.com/2017/09/07/business/equifax-cyberattack.html.
6. Mathews L. (2019). "Equifax Data Breach Impacts 143 Million Americans," Forbes Magazine, November 5, 2019. www.forbes.com/sites/leemathews/2017/09/07/equifax-data-breach-impacts-143-million-americans/#6a28ea86356f.

7. Schroeder P. (2018). "U.S. Charges Former Equifax Executive with Insider Trading," Reuter, March 14, 2028. https://www.reuters.com/article/us-equifax-cyber-sec/u-s-charges-former-equifax-executive-with-insider-trading-idUSKCN1GQ23W.
8. Riley M. (2017). "The Equifax Hack Has the Hallmarks of State-Sponsored Pros," Blomberg, April 30, 2020. https://www.bloomberg.com/news/features/2017-09-29/the-equifax-hack-has-all-the-hallmarks-of-state-sponsored-pros.
9. Ng A. (2018). "How the Equifax Hack Happened, and What Still Needs to Be Done," CNET, September 7, 2018. https://www.cnet.com/news/equifaxs-hack-one-year-later-a-look-back-at-how-it-happened-and-whats-changed/.
10. Yurieff K. (2017). "Equifax Data Breach: What You Need to Know," CNN, September 10, 2017. https://www.cnn.com/2019/07/25/us/equifax-700-million-settlement-data-breach-trnd/index.html.
11. Franceschi-Bicchierai L. (2017). "Equifax Knew of Security Flaws Months before It Was Hacked," April 30, 2020. https://www.vice.com/en_us/article/ne3bv7/equifax-breach-social-security-numbers-researcher-warning.
12. Krebson Security. (2020). "U.S. Charges 4 Chinese Military Officers in 2017 Equifax Hack." https://krebsonsecurity.com/2020/02/u-s-charges-4-chinese-military-officers-in-2017-equifax-hack/ (accessed February 10, 2020).

19 Office of Personnel Management Data

Abstract

This chapter is focused on the OPM incident that affected trust in people who thought their data in government data servers was safe and sound. It analyzes it from a human factors point of view.

INTRODUCTION

In 2015, the United States encountered one of its worst-ever data breaches. It was not on any financial or software organization but on a government organization, "Office of Personnel Management," commonly known as OPM. OPM is responsible for managing the federal government's civilian workforce. Its responsibilities include hiring and training employees, managing employee benefits and retirement plans, and conducting background checks on government employees and contractors. The OPM interagency incident response team discovered with high confidence that sensitive information, including the Social Security Numbers (SSNs) of 21.5 million individuals from the background investigation databases, was hacked. This includes 19.7 million individuals who applied for a background investigation and 1.8 million non-applicants, primarily spouses or co-habitants of applicants [1,2]. According to the OPM website, "some records also include findings from interviews conducted by background investigators, and approximately 5.6 million include fingerprints. Usernames and passwords that background investigation applicants used to fill out their background investigation forms were also stolen."

The OPM incident affected trust in people who thought their data in government data servers was safe and sound. It also illustrated the failure of security measures taken by the government. The breach exposed the vulnerabilities of the government information systems and raised concerns about the potential risks of centralizing sensitive information.

The OPM data breach significantly changed the government's cybersecurity policies and procedures. The government implemented new security measures to protect sensitive information better. Additionally, the breach prompted the creation of the Cybersecurity and Infrastructure Security Agency (CISA), a federal agency dedicated to protecting the nation's critical infrastructure from cyberattacks.

CASE DESCRIPTION

The OPM data breach consisted of two separate but linked attacks. It is unclear when the first attack occurred. The second attack happened on May 7, 2014, when the attackers posed as employees of KeyPoint Government Solutions, a subcontracting

DOI: 10.1201/9781003461739-25

company that provided background investigations, employment screening services, monitoring, and fraud prevention services. However, the second attack was not discovered until April 15, 2015 [1,2].

Attackers first breached the OPM networks in November 2013. The official congressional report claims that they were named "X1." They managed to get the IT system architecture information but failed to gain personal information. It seems that the OPM discovered that they were hacked first in March 2014. The cyberattackers had no personal data server access, so OPM didn't make this news public. OPM lets the attackers access the system and plans to monitor and cross-track them [3].

Unfortunately, another group, "X2," used stolen credentials from KeyPoint and established a backdoor on the OPM network. This backdoor remained untouched even when OPM performed the reset. Eventually, background investigation data was exfiltrated in August 2014.

The attackers were still inside the OPM servers without their knowledge, thus by completely studying their environment and architecture, they gained access to the server with personal records in October 2014.

During a hearing in front of the Senate Appropriations Subcommittee on Financial Services and General Government, OPM Director Katherine Archuleta told legislators that there was a direct line between the August breach of KeyPoint systems and the two intrusions of OPM servers detected in April [4].

A breach of KeyPoint Government Solutions—a contractor used by federal agencies to conduct background checks—gave hackers the credentials needed to access sensitive employee data held by the Office of Personnel Management.

Then, after around six months, they exfiltrated fingerprint data. Finally, one security personnel was alarmed when they noticed an unusual activity.

WHAT WE KNOW ABOUT THIS CASE

It is still vague if X1 and X2 were the same entity, but surely knowing that X1 already hacked OPM motivated and helped X2 to get pass through OPM's firewall. Also, they might be working in collaboration.

After X2 got into OPM's system, it was clear that OPM had not learned from its past mistakes. Primarily, OPM did not implement basic security measures like 2-factor authentication, which would have prevented X2 from breaking into the system just by the access of user credentials. Finally, in January 2015, OPM implemented two-factor authentication after being breached twice in this short time. Two-factor authentication requires the user to have a chip-enabled card or identity apart from just an ID and password. If it had been implemented before, millions of people's private information would have remained safe.

When X2 had OPM system access, it got root access through an Active Directory privilege escalation technique. Later, PlugX and Sakula were installed using root access. These are remote access tools. Attackers navigated around servers and extracted crucial information via these tools. They also got access to "jumpbox," the OPM admin server. It had the duty to log all the servers at OPM. This way, X2 had information on all the processes at other servers.

Finally, on April 15, 2015, a security person noticed something abnormal while examining OPM's networks. Later, it was discovered that OPM's complete infrastructure was connected to an unregistered website, "opmsecurity.org" through "mcutil.dll." mcutil.dll was something used by McAfee software, but OPM never used McAfee. Eventually, it was revealed that mcutil.dll cloaked remote access tools, and opmsecurity.org served as a controller server for X2. They also had hosted various similar sites, which at first glance seemed like official sites but were unregistered.

INFORMATION COMPROMISED

In the first breach in 2014, undisclosed sensitive information of people who applied for some jobs at OPM was extracted. However, this was never made official and is still denied [3].

The first large-scale breach of OPM's people's personal information happened in June 2015 when the training information, job applications, performance ratings, and information as vital as Social Security Numbers were breached. Information that was breached included:

- 1.1 Million Fingerprints: A staggering 19.7 million people who were part of OPM, applied for OPM, or were contractors lost their private information.
- 1.8 million records were indirectly related to the user whose background check occurred at some time.
- Personal information leaks included usernames, passwords, and details users filled out to complete their background checks.
- Information like illegal drug intake, medication, and financial failure was breached.

WHAT DID CYBERATTACKERS DO WITH THIS ATTACK?

There can be various ways in which attackers can use hacked information:

According to speculation, "China may be trying to build a giant database of federal employees" that will help them identify various U.S. officials and their roles [3].

There are also reports that the data obtained from the breaches can be used to access the systems by sending phishing emails.

Not only the United States but criminals can exploit these data to obtain information such as social security numbers and other personally identifiable information that can be used for identity theft.

The compromised data also includes payment details, which are the most critical and vulnerable data to exploit.

However, these data were not obtained for criminal purposes, as no trace of them was found underground or in the black markets. Even though the data was stolen for non-criminal purposes, it can still be used for criminal activities.

The fingerprint information can also be used to gain profit by trading it on the black market or else can be used to unveil the true identities of the undercover officials.

HUMAN FACTORS ANALYSIS

The OPM data breach highlights the human factors in this case. First, a human error prevented secure procedures such as stronger password requirements or multi-factor authentication from being implemented, which could have prevented such an incident. OPM came under intense scrutiny for its lax cybersecurity measures and failure to protect government employees' personal information adequately.

Miss Judgments

During the first attack, malware was noticed before hackers accessed private information. Then, the head of security, Katherine Archuleta, let them access OPM's servers so they could track back to attackers.

This resulted in attackers stacking up vital information, which later led to a second attack, which resulted in a data breach of over 21 million Americans.

In 2014, the Security team demanded the use of a security tool, Cyclace V by Cyclace, which was rejected, stating it did not pass security standards.

But later, in 2015, when OPM discovered they had been breached, they contracted Cyclave for the same tool, Cyclave V, to get out of the situation.

Also, OPM failed to implement basic security measures like two-factor authentication, which was the major cause of the second and the most deadly breach.

CONCLUSION

The OPM data breach exposed one of the highly confidential government organization's servers and proved that no data is completely immune to such attacks. This particular incident was linked to the Chinese Government. Few researchers thought that Chinese officials were directly involved in it. Also, two attackers, X1 and X2, were found to have their bases in China [1].

At the same time, this case highlighted the importance of basic security measures—such as two-factor authentication, network monitoring, system surveillance, and adequate training—that could have prevented such attacks.

REFERENCES

1. OPM. "Cybersecurity Resource Center Cybersecurity Incidentsm," OPM. https://nsarchive.gwu.edu/news/cybervault/2017-09-27/cyber-vault-opm-responsible-2015-hack-personal-info.
2. Boyd A. (2015). "Contractor Breach Gave Hackers Keys to OPM Data," Federal Times, June 23, 2015. https://www.federaltimes.com/smr/opm-data-breach/2015/06/23/contractor-breach-gave-hackers-keys-to-opm-data/.
3. Fruhlinger J. (2020). "The OPM Hack Explained: Bad Security Practices Meet China's Captain America," CSO, February 12, 2020. https://www.csoonline.com/article/3318238/the-opm-hack-explained-bad-security-practices-meet-chinas-captain-america.html.
4. Finklea K. et al. (2015). "Cyber Intrusion into U.S. Office of Personnel Management: In Brief," Congressional Research Service, CRS, July 7, 2015. https://sgp.fas.org/crs/natsec/R44111.pdf.

Part Seven

Financial Cybercrimes

- Emulators—Mobile Banking Frauds
- Google Store
- British Airways
- Target Data Breach
- TJ Maxx's Data Breach

20 Emulators–Mobile Banking Frauds

Abstract

This chapter focuses on the Emulators—Mobile Banking Frauds case, in which 16,000 individuals have online bank accounts, and raises many concerns about whether mobile banking is secure. It analyzes it from a human factors point of view.

INTRODUCTION

In 2020, a fraud operation was able to steal approximately 16,000 individuals' online bank accounts by using emulators to copy users' cellular devices and impersonating them to access their mobile bank accounts. This case raises many concerns about whether mobile banking is secure [1].

Millions of users use mobile banking in their daily lives. Unfortunately, cyber-attackers target users to hack and steal their bank information and use discretion. Online banking provides the convenience of performing simple tasks such as viewing their account balance and transactions, paying bills, and moving money to others and yourself.

Despite the improved security of mobile banking, there are still many pitfalls and cyber risks associated with using mobile banking. Cybercrimes continue to rise each year. Especially in the 21st century, the world is slowly adapting to big data and online transactions. The probability of cybercrime increases when users sign in and perform transactions on their mobile banking applications. According to Infosecurity, nearly 50% of mobile phones that have downloaded banking applications are at risk for fraud due to a lack of security. When using technology, both the application and users need to do their role to protect sensitive data [2].

In 2020, IBM Security Trusteer's mobile security research team uncovered a significant mobile banking fraud operation that successfully stole millions of dollars from financial institutions in Europe and the United States within a matter of days during each attack before being intercepted and stopped. This attack is conducted by organized crime groups using the infrastructure of mobile device emulators to set up thousands of spoofed devices that accessed thousands of compromised accounts. An emulator is typically a program that lets one run software from a completely different device on a computer; for example, an emulator can be used to play video games and run other operating systems like a Mac operating system on a Windows computer [1].

CASE DESCRIPTION

In this case, attackers used emulators to copy their victims' devices and steal millions of dollars from their online bank accounts.

DOI: 10.1201/9781003461739-27

The device identifiers employed enabled attackers to circumvent the banks' security measures. These identifiers were linked to each compromised account holder and included spoofed GPS locations. Subsequently, the attackers utilized the victims' identifiers obtained from compromised smartphones to access their bank accounts, masquerading as legitimate customers accessing accounts from new devices.

The scale of this operation has never been seen before. In some cases, over 20 emulators were used to spoof over 16,000 compromised devices and steal millions of dollars from users' online bank accounts [2,3].

The hackers were able to automate this attack. The automation of the attack allowed them to accelerate finding targets, stealing their information, and transferring the victim's money. This fraud is one of the biggest online banking frauds, with other means of cyberattacks that happen daily. It is becoming more challenging to trust devices and applications used daily. The case did not mention whether the victims could get their accounts back and any stolen funds. However, the case noted that the fraud operation "raises the usual security advice about using strong passwords" [2]. The attackers use these emulators to repeatedly access thousands of customer accounts and steal millions in just a few days in each case. After one spree, the attackers shut down the operation, wiped traces, and prepared for the next attack.

In each instance, a set of mobile device identifiers was used to spoof an actual account holder's device, likely ones that were previously infected by malware or collected via phishing pages. Then, using automation, scripting, and potentially access to a mobile malware botnet or phishing logs, the attackers, who have the victim's username and password, initiate and finalize fraudulent transactions at scale. In this automatic process, they could likely script the account balances of the compromised users and automate large numbers of fraudulent money transfers, carefully keeping them under amounts that trigger further review by the bank.

WHAT WE KNOW ABOUT THIS CASE

IBM researchers exposed one of the most extensive banking operation frauds [2]. The attack process was automated in this case, allowing the hackers to target larger groups.

Emulators that are legal to download and utilize are used to mimic the victim's mobile devices [4,5]. After each attack, a new emulated device was used. The attackers were using GPS spoofing to counterfeit the users' locations. For example, the attacker can be from anywhere and spoof the GPS to show the victim's location. This technique fakes the location. GPS spoofing is a form of cyberattack that involves manipulating GPS signals to provide false location and time information. This type of attack can have serious consequences, particularly when it comes to network synchronization and timing.

The attackers analyzed their techniques to improve their operations against other citizens. The attacks did not require a big team to perform. Also, device identifiers were used to avoid security protections set by the application.

INFORMATION COMPROMISED

The focus of these attacks was on banking. Therefore, information on their bank accounts was compromised. This information includes bank statements, loan information, account/routing number, and most importantly, access to control their funds. The protection can only go so far since hackers can imitate users' mobile devices and directly access their accounts. It was not mentioned whether the attackers sold the victim's usernames and passwords, but it was highlighted that the hackers-initiated funds transfer out of their accounts. The only information valuable to the hackers is the account and routing number. Although they do not have physical cards to withdraw money, they can still transfer the funds from the account [5,6].

WHAT ATTACKERS CAN DO WITH HACKED DATA?

When bank accounts get hacked, hackers can steal all the cash in the account by wiring it out. They will have access to users' routings and account numbers. Emulators are used to mimic the user's phone. Any text messages to confirm the user's identity will also be in their control. Another thing that hackers can do with this information is order checkbooks with their bank information. Once the checks are received, they can pay for purchases and write checks themselves.

Since the hackers have their accounts, they will access users' usernames and passwords. They will also be capable of running your credit and updating users' personal identification numbers to lock users' cards when they try to use them for their benefit. However, as mentioned earlier, most frauds that target personal finances are quickly caught and taken care of. Therefore, it is also essential to frequently check people's mobile banking statements and transactions for suspicion.

HUMAN FACTORS ANALYSIS

It is essential to have regular and frequent authentication whenever accessing applications on websites and mobile devices. Once the authentications fall into the attackers' hands, the user's unique identifiers, such as PINs, tokens, and other aspects of private information, are compromised.

Although PINS and passwords are commonly used for most phone applications as security measures, other means of authentication, such as biometrics and multi-factor authentication, can help prevent data from being stolen. For example, fingerprints and retinal scans are good security for authenticating the user. Since biometric data is complicated to change if the device is stolen, hacking biometric data will be extensive and tedious, and most likely, hackers will not attempt it. According to the case, it was mentioned that the hackers entered usernames and passwords into banking applications that ran on the emulators and mimicked the user's devices. Therefore, we can assume that they had no other security measures, such as biometric and multi-factor authentication.

A strong password is no longer effective if people do not have other forms of authentication. For example, the hackers used an emulator to mimic the victim's devices. Even if they have a multi-factor authentication setup that requires a password and PIN, the hacker can get the PIN since they are mimicking the user's device. The case also mentioned that they should use another form of authentication besides SMS.

It is also crucial that users be aware of phishing scams. Most people assume that phishing scams are only targeted at your work device, which is untrue. They can attack your work email and your email. Also, even calling you directly. Nowadays, most mobile devices have a feature that enables scam calls. The device will check if the person calling you is a person or a bot trying to collect your information when you pick up.

CONCLUSION

Cyberattackers often use emulators to manipulate mobile banking apps. Banks should invest in robust emulator detection mechanisms to identify and block transactions or account access from emulated devices, as well as other measures, such as biometric and multi-factor authentication. Until then, users should be aware that emulators can be used to conduct their own attacks to protect themselves from such attacks.

REFERENCES

1. Goodin D. (2018). "A Massive Fraud Operation Stole Millions from Online Bank Accounts," Wired, December 20, 2018. https://www.wired.com/story/massive-fraud-operation-stole-millions-online-bank-accounts/.
2. PT Security. (2020). "Vulnerabilities and Threats in Mobile Banking," PT Security, June 17, 2020. https://www.ptsecurity.com/ww-en/analytics/vulnerabilities-mobile-banks-2020/.
3. Johnson D. (2020). "Emulators Can Turn Your PC into a Mac, Let You Play Games from Any Era, and More – Here's What You Should Know about the Potential Benefits and Risks of Using One," Business Insider, October 16, 2020. https://www.businessinsider.com/what-is-an-emulator (accessed May 2, 2021).
4. Bendovschi A. (2015). Cyber-Attacks – Trends, Patterns, and Security Countermeasures. *Procedia Economics and Finance*. Vol. 28, pp. 24–31.
5. Dan Goodin A. T. (2020). "A Massive Fraud Operation Stole Millions from Online Bank Accounts," Wired, December 18, 2020. https://www.wired.com/story/massive-fraud-operation-stole-millions-online-bank-accounts/ (accessed April 27, 2021).
6. Coker J. (2020). "Widespread Security Vulnerabilities in Mobile Banking Apps," Infosecurity Magazine, June 17, 2020. https://www.infosecurity-magazine.com/news/security-vulnerabilities-mobile/#:~:text=Half%20of%20mobile%20banks%20are,sensitive%20data%20and%20commit%20fraud (accessed May 1, 2021).

21 Google Store

Abstract

This chapter is focused on the Google Store case that tricked 560,000 people into downloading a gaming app with an extra malware app to gain full network access to a user's device. It analyzes it from a human factors point of view.

INTRODUCTION

Google's (Android Play Store) security breach surfaced in November 2018 and has affected more than half a million Google Play Store users. In this case, the hacker used a sophisticated process to inject malware onto the user's device or network. More than 560,000 were tricked into downloading the gaming app, which included a mix of luxury car and truck simulation apps and required users to download an extra malware app to gain full network access to a user's device [1].

Malware can steal, delete, encrypt, or alter data from the device and the network within which the affected device lies. Malware can also hijack core computer functions and spy on the network without user knowledge or permission.

Google's Play Store offers millions of Android OS applications. Unfortunately, several Android-OS-related security breaches have proved that the Play Store is becoming notorious for hosting fake and infected applications [2].

Infected applications can be harmless adware (that does not utilize the device's computational power) or notorious malware that can adversely affect user data.

Previously, malicious banking applications on the Play Store were imitated from six leading banks in the United Kingdom, Australia, Switzerland, New Zealand, and Poland. In September 2017, ESET, an antivirus development organization, found out about such applications. In addition, ESET realized that these fake (imitations) apps collected the user login credentials and several other sensitive financial artifacts and used them for monetary benefit or to blackmail users [3].

CASE DESCRIPTION

In November 2018, an ESET security firm researcher, Lukas Stefanko, discovered a few car racing games on the Play Store. The games were all designed by the same developer, Luiz Pinto. The peculiar fact that caught his eye was that all these 13 gaming applications required users to download and update the application from a very unsafe source. Moreover, the manual requirement for downloading separate files for the game made Lukas even more suspicious, and they tweeted that they downloaded and loaded this app [4]. He then tweeted his findings, which made people realize there was a security issue. Google was fast enough to respond to this tweet and remove all these applications from the Play Store. A senior security expert at Google,

DOI: 10.1201/9781003461739-28

Scott Westover, claimed in response to Lukas's tweet, "Providing a safe and secure experience for our users is our top priority. We appreciate the researcher's report and their efforts to help make Google Play more secure. The apps violated our policies and have been removed from the Play Store" [5].

However, this measure needed to be more timely since these applications already had over half a million users. Moreover, one of the appalling facts about Lukas's tweet was that two of these games were on the top trending list of the Google Play Store, with more than half a million downloads in almost a week. The following section discusses how the malware is injected and its consequences [5,6].

WHAT WE KNOW ABOUT THIS CASE

Whenever a user downloads the application, the installation procedure looks equivalent to other typical applications. However, when the installation process is finished, the application deletes the game icon/logo, which users believe is why the installation process failed. The applications have no genuine usefulness and serve only as a cover for malware to be downloaded.

It is the Android application package file format used by the Android operating system and several other Android-based operating systems to distribute and install mobile apps, mobile games, and middleware.

When the users were installing the game, it prompted them to download an extra Android Package (APK), a file format used by the Android operating system, which was the malware. The malware mimics a Game Center application and asks for the user's consent to be downloaded and set up on the device. The idea behind deleting the game icon after installation is to create an illusion in the users' minds that the previous installation might not have been successful because of the lack of the game icon in their application drawer and, hence, the second installation.

In this way, the hacker takes advantage of the user's mindset, and hence, the malware is injected into the user's device. Upon complete injection, when the device is unlocked, it is attacked by a surge of adware, making it noticeably slow down. This surge of adware is followed by malware taking complete control over the device's data and network.

This tweet gained immense attention and compelled several anti-malware researchers working with reputed organizations to determine the malware family to which this malware belongs. But unfortunately, most advanced software platforms labeled this a "Hidden-ad Trojans" [7,8].

INFORMATION COMPROMISED

The second application that the users download, which the hacker calls the "Game Center," lets the hacker gain full network access to the user's device. Once this happens, the compromised information includes the total network traffic going in and out of the host device's network.

The application asked for permission from the OS as soon as the device started up or was unlocked. Then, the device was started up or unlocked. Users should have noted that hackers gained remote "full access" to the host device's network traffic;

it is not just the host device data at risk but each device's data part of this network. Moreover, the incoming data onto this network is compromised, making the situation even more grave.

WHAT CYBERATTACKERS CAN DO WITH HACKED DATA

The hacker can use the data and encrypt it to ask for ransom from the users. In addition, misuse of essential data poses a serious threat. The hacker can misuse the compromised data for monetary or other benefits. Moreover, with unprecedented access to the network data, the hacker can steal important data that belongs not just to the one who downloaded the malicious application but also to each device on the network and to each device that is sending data to this network. Network data can be used by hackers for surveillance as well.

HUMAN FACTORS ANALYSIS

It's astonishing how aware the hacker was from a human–computer interaction viewpoint. He analyzed that people can and will consider redownloading an application hastily if its icon does not exist in their app drawer after the first installation. The hacker designed the application icons, screenshots, description videos, and all the information displayed on the Play Store in such an immaculate and beautiful manner that the users were tempted to try out the game. The hacker used some human behavior vulnerability when designing the UI.

CONCLUSION

In 2017, Google removed approximately 700,000 apps from the Play Store and made it more challenging for hackers to upload such applications [9]. To enhance security standards, Google added a metadata string to every Android APK file to ensure applications are secure and verified. It is crucial to raise awareness about these threats among users and educate them on preventive measures. Additionally, it is essential to update people's devices regularly, encrypt and back up the data, and carefully review the permissions granted to each application.

REFERENCES

1. PriceE.(2018)."MoreThan500,000PeopleDownloadedTheseMalware-InfectedAndroid Apps," Fortune, November 26, 2018. https://fortune.com/2018/11/26/google-play-malware-apps/?utm_source=twitter.com&utm_medium=social&xid=soc_socialflow_twitter_FORTUNE&utm_campaign=fortunemagazine.
2. Doffman Z. (2024). "New Report Warns Android Users about Dangers of 'Copycat Apps'," Forbes, January 13, 2024. https://www.forbes.com/sites/zakdoffman/2024/01/13/urgent-new-warning-suddenly-issued-for-millions-of-google-android-users/.
3. Paulina Duran P. (2018). "Two Australian Banks among Six Targeted by Fake Apps: Security Firm," Yahoo News, September 19, 2018. https://www.yahoo.com/news/two-australian-banks-among-six-targeted-fake-apps-043351233--sector.html?guccounter=1&guce_referrer=aHR0cHM6Ly93d3cuZ29vZ2xlLmNvbS8&

guce_referrer_sig=AQAAABx7BgBirWkF3WQUFvJCV_h3hTKtSxSq0eREDz-9tZj0gAU1MIdagLx1KvYJi1wY2zi_ceD7zsJ-FdeoqqDZAO950OaG_kVbRyYZr-Fpb-2DHa-CsHOZD6co5T9bDTJ0aeKQHpkxzLVr389By0aQwqZKfTYscCVy5z5ny-hDIF1YYLl.

4. Brewster Th. (2018). "500,000 Duped into Downloading Android Malware Posing as Driving Games on Google Play," Forbes, November 19, 2018. https://www.forbes.com/sites/thomasbrewster/2018/11/19/500000-duped-into-downloading-android-malware-posing-as-driving-games-on-google-play/?sh=6e9776e91c2b.
5. Villas-Boas A. (2018). "More Than 500,000 People Downloaded Games on the Google Play Store That Were Infected with Nasty Malware—Here Are the 13 Apps Affected," Insider, December 1, 2018. https://www.businessinsider.com/google-play-store-game-apps-removed-malware-2018-11.
6. Stefanko L. (2018). "Don't Install These Apps from Google Play – It's Malware," Lukas Stefanko Twitter Account, November 19, 2018. https://twitter.com/LukasStefanko/status/1064507886896844800.
7. Whittaker Z. (2018). "Half a Million Android Users Tricked into Downloading Malware from Google Play," TechCrunch, November 20, 2018. https://techcrunch.com/2018/11/20/half-a-million-android-users-tricked-into-downloading-malware-from-google-play/.
8. Lekach S. (2018). "Driving Apps Pulled from Google Play for Reportedly Installing Android Malware over 560,000 Times," Nashable, November 20, 2018. https://mashable.com/article/apps-malware-google-play-store-breach.
9. Chris Welch Ch. (2018). "Google Took Down over 700,000 Bad Android Apps in 2017/ That's 70 Percent More Than 2016's Removals," The Verge, January 30, 2018. https://www.theverge.com/2018/1/30/16951996/google-android-apps-removed-security-2017.

22 British Airways

Abstract

This chapter is focused on the case of British Airways, in which customer data was compromised from the British Airways website and mobile application. The hackers gained access to the customers' confidential data. It analyzes it from a human factors point of view.

INTRODUCTION

Software technologies are growing rapidly over the last few decades. People's day-to-day transactions are carried out online using software technologies. The transactions, like booking a ticket, transferring money, checking bank statements, etc., are carried out online using the software. This software uses and records a lot of user data to carry out such transactions. The user data may include credit or debit card information, passport information, address proofs, and other private information. The user data is very confidential, and hackers try to infiltrate the software to access the data. Cyberhackers infiltrated the software mostly because of the software's bad security standards. The breached user data is then used for many wrongdoings, thereby causing a lot of mental and economic stress to the users.

Air travel is one of the fastest ways to travel from one place to another. Because of time efficiency, air travel has been very common for the past few decades. Due to the advancement in software technology, people are booking air tickets online from their homes with a few clicks. All the airways companies have a website or mobile application to book the tickets. The user must provide basic information like name, email, address, travel information, etc. to book the ticket. Once the user has provided the required travel location and basic information, one is asked to pay using the credit or debit card information. The air ticket is confirmed on the user's name once the payment is confirmed. The process of booking the air ticket is very simple and efficient, with very little hassle. Since the user is providing the basic information and payment information to the airline vendor, the vendor needs to protect the user data by implementing strong security standards. It is the airline's responsibility to protect and make sure that user data is kept secure at whatever cost.

Hackers constantly try to break into airline systems to gain access to user data. There have been many incidents of data breaches in the airline industry. The British Airways customer's data was breached between August 21 and September 5, 2018.

Close to 380,000 customer data was compromised from the British Airways website and mobile application between the two. The hackers gained access to the customers' confidential data.

DOI: 10.1201/9781003461739-29

General Data Protection Regulation (GDPR), a European Union regulation on information privacy that investigates the regulation of protecting customer data, imposed a fine of little more than 180 million Euros, making it the biggest data breach fine in history. As per GDPR, the penalized company can be charged up to 4% of the company's yearly earnings. In the case of British Airways, this fine accounted for up to 2% of their yearly earnings. British Airways had even challenged this fine. After this breach, British Airways has taken appropriate actions to bolster its security standards to avoid any similar data breach attacks in the future [1]

CASE DESCRIPTION

British Airways did not officially announce how the hackers carried out the data breach. There is not enough information here about the breach, only speculation. However, the Information Commissioner Office (ICO) proposed an idea about how the breach may have been carried out: the customers were taken to a malicious site from the British Airways website ba.com during the transaction of booking the ticket. On this malicious website, the user's confidential details, including personal data, were collected by hackers. British Airways also gave assurance to the affected customers [2].

WHAT WE KNOW ABOUT THIS CASE

According to the experts, an insider inserted malicious code into the British Airways codebase. This malicious script collected and sent the user's personal and payment information to the hackers' server.

Experts even mentioned the possibility of malicious code in the third-party payment vendor code base. British Airways uses such vendors to complete customer payment transactions. So, there is a possibility of a lack of security measures for these vendors, thereby causing a security breach. Also, there is a possibility of the transaction being hijacked between the British Airways software and third-party payment vendors. The hackers could have interpreted the transactions and got hold of confidential information. Since there is no statement from the Airways, all the theories are just speculation, and no theory can be proved correct [3].

INFORMATION COMPROMISED

The data compromised in this breach included customer names, emails, billing addresses, payment card information like card number and CVV number, and even the customers' travel plans. Customer numbers were also breached. However, this breach did not compromise personal identity documents like passport numbers.

WHAT CYBERATTACKERS CAN DO WITH HACKED DATA

In the British Airways breach, all the private and confidential data of the customer is available to the cyberattackers. Attackers can use data for:

Identity Thief: Attackers can use the customer's real data to create accounts, thereby giving the impression that the customers have created these accounts themselves.

They can even use this data for illegal transactions, such as buying illegal items. Basically, the customer's identity can be used without even letting one know. Hackers also have access to the payment card information, including the customer's CVV, and can use the card information to purchase anything from anywhere. The hackers can carry out illicit transactions without raising alarms, as these transactions will look real until the affected customer notices it.

The customers affected by this breach can incur identity or money theft, thereby causing mental trauma to the customer [4].

HUMAN FACTORS ANALYSIS

The details of how cyberattackers write well-written code are defined by how good the security standard is. Code security standards vary as humans write the code, and the factors the coder evaluates before writing the code may vary from person to person. In the case of a British Airways data breach, a malicious piece of JavaScript may be inserted in the third-party payment vendor or their code base.

The hackers were able to manipulate the scripts and change them. The changes in the script are traceable if they are monitored regularly and high-security standards are maintained. So, the written code and the security standards are prone to human error. In the case of British Airways, a malicious website was called from their website. The engineers should have kept track of the website from their site. The engineers should have kept track of the server site scripts. By doing this, the breach could have been avoided.

CONCLUSION

The British Airways data breach in September 2018 was a significant cyberattack, affecting around 380,000 transactions and exposing customers' personal and financial details. This attack illustrates how important management behavior is. British Airways took over two weeks to detect the breach, highlighting the need for robust monitoring systems to identify and respond to security incidents quickly.

Another critical factor was Web Application Security. Since the breach was linked to malicious code on British Airways' website, it indicates the need for secure coding practices, regular code reviews, and application security testing. Also, sensitive customer data should have been encrypted at rest and in transit.

The breach occurred after the GDPR came into effect; thus, British Airways should have complied with it to avoid a fine. This is another human agent-related factor.

Another human-related factor is that the breach involved a third-party vulnerability. British Airways should have ensured that its vendors and partners adhered to strict security standards and conducted regular audits.

The British Airways data breach illustrates how human factors play an important role in cybersecurity.

REFERENCES

1. Whittaker Z. (2018). "British Airways Customer Data Stolen in Data Breach," TechCrunch, September 6, 2018. https://techcrunch.com/2018/09/06/british-airways-customer-data-stolen-in-data-breach/.
2. BBC. (2019). "British Airways Faces Record £183m Fine for Data Breach," BBC, July 8, 2019. https://www.bbc.com/news/business-48905907.
3. BBC. (2018). "British Airways Breach: How Did Hackers Get in?" BBC, September 7, 2018. https://www.bbc.com/news/technology-45446529.
4. Calder S. (2019). "What Is the British Airways Data Breach and How Does It Affect Passengers?" Independent, July 8, 2019. https://www.independent.co.uk/travel/news-and-advice/british-airways-data-breach-pr ivacy-details-leak-iag-cathay-pacific-information-commissioner-a8993331.html.

23 Target Data Breach

Abstract

This chapter reviews Target's case and the causes of the widespread and possible contributory factor-like human factors.

INTRODUCTION

Target Corporation is an American department store chain. It is a well-known retailer that offers customers various products and services, from electronics to clothing and groceries.

Target's nationwide data breach took place in November 2013. The Target data breach incident offers a classic example of attackers exploiting common vulnerabilities and accessing information by exploiting human behavior. In addition, the cyberattacks of 2013 affected millions of people, and more than 2,000 Target stores were attacked at different geographical locations.

Retail businesses are a common target for cyberattackers. They often store vast amounts of consumers' personal information, such as names, credit card information, social security numbers, email addresses, and passwords (passwords can be common across accounts).

This chapter reviews Target's case and the causes of the widespread and possible contributory factor-like human factors.

CASE DESCRIPTION

As a result of this data breach, the personal data and financial information of Target Corporation customers and 40 million debit and card numbers were stolen [1]. In addition, the personal data of 70 million customers, including addresses and phone numbers, was compromised. Target officials disclosed this security breach during the Thanksgiving season in December 2013. The data breach continued between November 27 and December 15, 2013, from countrywide Target retail stores where customers had shopped. According to Krebs on Security, the hacked data of credit and debit cards were made available on the underground website for $20 to $100 per card [2–4]. Target took no remedial action despite information on the data breach given by the U.S. Department of Justice in mid-December and notifications of an infringement by internal sources [3].

The attackers first targeted third-party vendors and compromised one to steal credentials for Target's system. After using the stolen credentials, the hackers gained access to Target's system. They accessed their database and installed malware on their system. They captured the consumers' names and contact information, including phone numbers, email addresses, street addresses, credit/debit card information, credit card verification codes, and other sensitive data.

DOI: 10.1201/9781003461739-30

After discovering the data breach, Target provided a free credit monitoring service for the customers' payment accounts affected by the incident. As part of a $10 million lawsuit settlement reached in 2015, Target also agreed to pay up to $10,000 to customers with proof of their losses due to the data breach. Target also agreed to pay an $18.5 million multistate settlement for affecting over 41 million customers' payment card information [5].

Target publicly acknowledged this leak on December 19, 2013, after the hack was initially uncovered on December 18, 2013 by Brian Krebs, an investigator specializing in news about internet security [2]. During the testimony at U.S. Congress, the officials from Target said they were unaware of the incident until December 12, when the Department of Justice contacted them [4,6].

WHAT WE KNOW ABOUT THIS CASE

The data breach incident happened in the following stages:

1. Initial Survey: The attackers performed reconnaissance on Target's network to gain information about Target's systems' architecture. One of the papers describes that such information was readily available on the internet at that time. Studying this information helped the attackers to understand how Target's infrastructure and Point-of-Sales system worked.
2. Compromise Third-Party Vendor: Next, the attackers searched the Internet for a list of third-party vendors. They found information on Target's website about all their vendors, how they interacted with Target's system, and other important information. The attackers then entered Target's network by compromising a third-party vendor. They started trying to attack these vendors by sending out phishing emails embedded with fake links, leading to malware.

   An employee of Fazio Mechanical Services, Inc. (U.S.-based heating, ventilation, and air-conditioning company) was tricked into clicking this fraudulent link from that phishing email. That led to downloading malware, a variant of Trojan malware, on the employee's system. The attackers gained that employee's login credentials for Target's portal and used them to get a foothold inside Target's system.
3. Attack on Target's POS Systems: The attackers then were able to run the malware called "Trojan. POSRAM" on the POS systems. This malware scraped the POS's RAM to steal the credit/debit card information as the cards were swiped on the POS devices and sent that information over to a dump server at regular intervals. After the allotted time had elapsed, the malware indicated to the attackers that the stolen data was successfully dumped on the dump server by sending an Internet Control Message Protocol (ICMP) ping packet to the remote server. ICMP is a network layer protocol used by network devices to diagnose network communication issues. The hackers then moved the stolen data to their servers.

In this way, the hackers successfully carried out this attack.

The Target data breach could be attributed to multiple factors:

1. Vendors were highly susceptible to phishing attacks.
2. Lack of network segregation.
3. Point-of-sale systems were not protected from malware.
4. The data protection system used by Target was not effective.

The breach occurred at the third-party vendor premises, namely Fazio Mechanical Services. The hackers used Trojans to access the Target systems like billing and business networks. Target's security system was weak at third-party premises; therefore, hackers could access Target's network through the Fazio Mechanical terminal. Target did not put sufficient protections on third-party terminals to secure its network, leading to a massive breach of data [4,7].

The third-party terminal could access customers' sensitive data without restriction, which indicates that Target's network was operating without protection at different levels. The payment system of Target's network was not sufficiently protected from third-party access, and the network was not properly segmented and protected.

Third-Party terminal: The improperly segmented network without protection allowed hackers to access Target's network and steal customers' data.

The next step for the hackers to steal the information was installing malware at the point of sale. Since the network did not have enough protection against malware for sensitive data, the hackers had unhindered access to sensitive data. Through this route, there was easy access to in-store point-of-sale systems, making it easy to steal the encrypted information directly from the magnetic strip of the customers' cards. The hackers also took over three FTP servers of Target's internal network, and this stolen information was stored over these servers. This highly encrypted information is then sent to drop sites in Miami and Brazil. The ultimate step is to upload this card information to the black market to sell individual cards. Other personal details, such as the names, addresses, and phone numbers of the purchasers, were also leaked, leading to identity thefts on a mass level [4,5].

The Internal Control of any retail company is a vital part of the company's security. They play an important role in managing risk and preventing fraud in a company. The policies and procedures to safeguard knowledge, including stored data on a network, economic reporting reliability, and strict adherence to regulations and law, are to be adopted by management for their company's safe and secure working. The hackers who intended to target various American retailers in 2013 discovered that Target's internal control system was deficient in many aspects and hence would be a more accessible aim for a cybercrud. They could trace that Target's network lacked motion detectors and virtual walls, which form the backbone of every security system. This lack of a basic security system made it a laid-back job for criminals to use Target's third-party vendors to wreak havoc on Target. The whole breach process was easy, requiring only basic hacking knowledge. Still, the only sophisticated component of this attack was the malware used to hack Target's POS systems, which required an advanced skill set.

Target, however, was prepared for a similar situation. Anticipating that a breach like this might occur, the corporation installed malware detection software through a security firm named FireEye. FireEye had a high-profile clientele, such as the CIA and the Pentagon. The firm, situated in Bangalore, India, continuously monitored Target's network and would transfer this information to Target's Internal control team in Minneapolis [8].

At Thanksgiving, when the hackers installed the malware on Target's servers to maneuver the stolen information, this data was routed through the United States to various drop locations in Russia and Brazil. The FireEye team immediately picked up this suspicious activity on the servers, and a notification about the same was also conveyed to the security team of Target in Minneapolis [9]. However, despite an alert from the anti-malware company, Target's internal control team did not consider acting upon the alert for unknown reasons and thus jeopardized important customer information.

INFORMATION COMPROMISED

Information Compromised in the Data Breach

The following information was compromised during the Target data breach:

- Full names
- Phone numbers
- Email addresses
- Payment card account information
- Credit card verification codes and other sensitive data

WHAT CYBERATTACKERS CAN DO WITH HACKED DATA

Two main types of information were compromised during the breach it can be classified:

- Customers' personal information such as names and contact information [9,10].
- Customers' payment account information such as credit card/debit card numbers and credit card verification codes.

Hackers can sell customers' personal information on the dark web. Such information can also be used to impersonate clients and commit identity fraud. It can also be used to perform financial fraud; for example, banks often use this information to verify a person's identity before authorizing a transaction.

The second type of information can be used by hackers to perform fraudulent financial actions and gain money. Again, this information can be sold on the dark web for a very high price.

HUMAN FACTORS ANALYSIS

From the human factors point of view, this case offers interesting insights into human–computer interaction. One of the key insights is the general need for more awareness about security practices in the work environment. This is evident from the fact that the third-party vendor was not alert when he clicked on the phishing link sent by the attackers. Extensive security training in the workplace would have easily prevented this mistake. It is important to understand that investing in such training occasionally goes a long way in strengthening the security of business data.

Another important lesson is that there exists a need for seriousness about security practices in the work environment at both the third-party vendor and Target's end. For example, Fazio Mechanical acknowledged using a free version of antivirus software that could not detect the specific malware the attackers used. They could easily have used a paid version capable of detecting that and much more dangerous malware for better security; however, they did not. Also, Target could have made it mandatory for third-party vendors to secure their systems or forced them to employ practices such as two-factor authentication, but they did not. Also, Target's security alarms sent warnings at intervals during the attack. However, they were not heeded by the Target system experts, which led to the continuation of the attack. This needs to prioritize security at the workplace costs dearly to both businesses.

Another important learning is to fix the system vulnerabilities promptly. Target could have fixed the vulnerabilities, which would have prevented the lateral movement of attackers after gaining a foothold in Target's system. This would have ensured that even if the attackers gained an entry, the blast radius of their attack would have been minimal. In addition, regular security audits would have exposed the system's vulnerabilities. Thus, a good security process could have mitigated the effects of the data breach by a considerable margin.

CONCLUSION

The Target data breach in 2013 was one of the most significant security incidents, affecting around 40 million credit and debit card accounts and 70 million customer records. Several human-related factors were identified in the case. First, the issue was the third-party vulnerability. The IT professionals at Target should have ensured that its vendors and partners adhered to strict security standards and conducted regular audits.

The second factor might be the failure to fix the system vulnerabilities. Fixing the vulnerabilities would have prevented attackers from lateral movement after gaining a foothold in Target's system. Other factors might be a lack of proper training of IT professionals to monitor threat detection and respond to potential threats [11,12].

REFERENCES

1. Clerix K. (2014). "The Target Breach, By the Numbers," Krebson Security, May 6, 2014. https://krebsonsecurity.com/2014/05/the-target-breach-by-the-numbers/.
2. "Cards Stolen in Target Breach Flood Underground Markets," KerbsSecurity.Com, December 20, 2013. https://krebsonsecurity.com/2013/12/cards-stolen-in-target-breach-flood-underground-markets/.
3. McMullen D. A., Sanchez M. H. et al. (2016). Target Security: A Case Study of How Hackers Hit the Jackpot at the Expense of Customers. *Review of Business & Finance Studies*. Vol. 7, No. 2, pp. 41–50.

4. Radichel T. (2014). "Case Study: Critical Controls That Could Have Prevented Target Breachat," Sans Institute, September 12, 2014. https://www.sans.org/reading- room/whitepapers/casestudies/case-study-critical-controls-prevented-target-breach-35412.
5. McCoy K. (2017). "Target to Pay $18.5M for 2013 Data Breach That Affected 41 Million Consumers," USA Today, May 23, 2017. https://www.usatoday.com/story/money/2017/05/23/target-pay-185m-2013-data-breach-affected-consumers/102063932/.
6. Plachkinova M., Maurer C. (2018). Teaching Case: Security Breach at Target. *Journal of Information Systems Education*. Vol. 29, No. 1, pp. 11–20.
7. United State Senate. (2014). "A 'Kill Chain' Analysis of the 2013 Target Data Breach," Committee on Commerce, and Transportation, March 26, 2014. https://dd80b675424c132b90b3-e48385e382d2e5d17821a5e1d8e4c86b.ssl.cf1.rackcdn.com/files/external/Target_Kill_Chain_Analysis_FINAL.pdf.
8. Xiaokui S., Ke T., Ciambrone A., Yao D. (2017). "Breaking the Target: An Analysis of Target Data Breach and Lessons Learned," Cornel University, January 18, 2017. https://arxiv.org/abs/1701.04940.
9. Kassner M. (2015). "Anatomy of the Target Data Breach: Missed Opportunities and Lessons Learned," ZDNET, February 2, 2015. https://www.zdnet.com/article/anatomy-of-the-target-data-breach-missed-opportunities-and-lessons-learned.
10. Clark M. (2014). "Timeline of Target's Data Breach and Aftermath: How Cybertheft," IB Times, May 5, 2014. https://www.ibtimes.com/timeline-targets-data-breach-aftermath-how-cybertheft-snowballed-giant-retailer-1580056.
11. CEI Blog. (2013). "Cyber Security: Target's 2013 Data Breach," CEI, Blog, September 21, 2015. https://www.cei.com/about-cei/media-room/blog/cyber-security-targets-2013-data-breach/.
12. Shu X., Tian K., Ciambrone A., Yao, D. D. (2017). "Breaking the Target: An Analysis of Target Data Breach and Lessons Learned," ArXiv, abs/1701.04940. https://arxiv.org/pdf/1701.04940.

24 TJ Maxx's Data Breach

Abstract

This chapter reviews the TJMAXX case. This chapter analyzes this case from a human factors point of view since the human agents' behavior patterns that enable this attack to be successful are interesting.

INTRODUCTION

TJX Companies Inc. is a home fashion retailer headquartered in the United States. TJ Maxx store chain has more than 1,000 stores in the United States. In 2007, the attackers stole millions of credit card numbers from TJX Inc. by intercepting wireless transfers of customer information at two Miami-area Marshall stores. The company did not reveal the data breach for two months, then it finally offered more details about the extent of the compromise in filings with the U.S. Securities and Exchange Commission and acknowledged that 45.6 million credit and debit card numbers were stolen from one of its systems over more than 18 months by an unknown number of attackers.

According to reports, TJMAXX had collected too much personal information, kept it too long, and relied on weak encryption technology to protect it [1,2].

Although the TJX case is not very recent, the human agents' behavior patterns that enable this attack to be successful are still interesting and worth attention and analysis.

This chapter reviews this case.

CASE DESCRIPTION

TJX's parent company's wireless network was secured with the weakest forms of security for wireless Local Area Networks (LANs). Consequently, attackers were able to break into the system and steal millions of credit card numbers from the second half of 2005 and throughout 2006.

According to the reports, attackers cracked the Wired Equivalent Privacy (WEP) encryption protocol used to transmit data between price-checking devices, cash registers, and computers at a store in Minnesota. The intruders then collected information submitted by employees logging on to the company's central database in Massachusetts, stealing usernames and passwords. Then, the attackers set up their accounts on TJX's system and, over 18 months, collected transaction data, including credit card numbers, into approximately 100 large files. Transaction data sent to banks, which TJX unencrypted, is also believed to have been intercepted by the hackers.

Interestingly, the attackers even left encrypted messages on the TJX network to tell each other which files had been copied [3].

DOI: 10.1201/9781003461739-31

WHAT WE KNOW ABOUT THIS CASE

According to the reports in 2005, on multiple occasions, attackers sat outside of a Marshall store in St. Paul, Minnesota, and pointed a telescope-shaped antenna toward the store, capturing wireless transactions inside the store as it was being broadcasted through the company's wireless network. Thus, the attackers captured the traffic through the networks making sure they were invisible and that no one knew they were there.

After capturing information for several days, the attackers broke the store's WEP security code and stole credit card and bank account information. It is reported that the attackers also gained access to the system and compromised the TJX headquarters' corporate networks in Framingham, Massachusetts. It seems that they also had access to company and customer information from the company's centralized corporate database [4].

In October 2006, TJX retail stores were found to be subject to a data breach. Later, the company said that the intruder initially gained access to the system via the wireless local area networks (WLANs) at two stores in St. Paul, Minnesota [5].

On December 17, 2006, TJX learned that suspicious software had been detected in its computer systems and the computer systems had been breached and that an intruder remained on the corporate systems. At that time, it was unclear as to who was responsible for the intrusion or whether there were one or multiple separate intrusions. Internally, TJX did not find evidence that any of its employees were involved in this incident. The company later confirmed its suspicions that the intruder initially gained access to the system via the WLANs at two stores in St. Paul, Minnesota [5].

After analyzing the evidence, TJX notified various U.S. law enforcement agencies of the suspected intrusion.

With the agreement of law enforcement, the company informed its contracting banks, credit card, debit card, and check-processing companies of the suspected intrusion. TJX determined that one of its systems had also accessed customer information during the computer intrusion [5].

The attackers stole 45.7 million credit and debit cards of shoppers' data from T.J. Max and Marshalls stores. The stolen data were from transactions beginning in January 2003 and ending November 23 of that year [6].

INFORMATION COMPROMISED

According to the reports, TJX stored customer personal information in its Framingham system. The information received from its stores in the United States, Puerto Rico, and Canada pertained to merchandise returns without receipts and some check transactions. The personal information consisted of driver's license numbers and identification (ID) numbers (such as military and state ID, in some cases including social security numbers), together with the names and addresses of the customers who had returned goods. Since April 7, 2004, the practice was to encrypt the information before storing it. Actual characters were substituted by encryption using an encryption algorithm provided by the software vendor.

TJX assured its payment card customers that their names and addresses were not included with the payment card data believed stolen for any period because TJX did not process or store that information on either the Framingham or the Watford system in connection with payment card transactions. It also assured its customers that by April 3, 2006, the Framingham system was masking payment card PINs and some portions of check transaction information.

For transactions after April 7, 2004, the Framingham system encrypts all payment card and check transaction information. Concerning the Watford system, masking and encryption practices had been implemented at various points in time for various portions of the payment card data.

That number eclipses the 40 million records compromised in the mid-2005 breach at CardSystems Solutions and makes the TJX compromise the worst ever involving personal data loss [7].

WHAT CYBERATTACKERS CAN DO WITH HACKED DATA

With a driver's license, state ID, and social security number, individuals can use these forms of identification for various purposes, including creating a fake driver's license required to operate a motor vehicle legally. A state ID can also be used for various purposes, such as voting and opening a bank account. More importantly, the social security number can be used to apply for social security benefits, track earnings for tax purposes, and obtain employment. A driver's license or state ID and a social security number can be used to apply for credit, loans, and other financial transactions.

HUMAN FACTORS ANALYSIS

Several areas in this case are related to human factors, including errors by IT professionals, management, and employees.

IT PROFESSIONALS ERRORS

Attackers cracked the Wired Equivalent Privacy (WEP) encryption protocol used to transmit data without strong encryption. When the attack happened, it was already known that WEP is obsolete and not secure. TJX should have implemented WPA encryption technology.

TJX should have also actively monitored and tested its WLAN security. Knowing that the wireless networks collect and transmit data via radio waves, information about purchases and returns can be shared between cash registers and store computers. Antennas can intercept wireless transmissions, and high-power models can sometimes intercept wireless traffic from miles away. The investigators revealed that TJX had two years to convert from WEP to more sophisticated Wi-Fi Protected Access, and it did not do so.

TJX did not have a process for updating all its critical software components and applying any security patches released by the software vendors.

Management Errors

There are several issues with management behavior.

As a retailer that accepts credit cards, TJX was not compliant with PCI Security Standards. As a corrective control, TJX should actively implement the PCI standards.

TJX collected too much personal information, kept it too long, and relied on weak security encryption [1,8,9].

TJX retained customer data years after it should have been purged, including driver's license numbers collected when customers returned merchandise without receipts. Some of the stolen information was from transactions concluded as long as 2002.

TJX management behavior is also another part of human-related factor by not informing the customer.

CONCLUSION

The TJX Companies Inc. data breach was one of the largest retail cybersecurity incidents in history. It affected millions of customers and significantly affected Target and the broader cybersecurity landscape.

Some key lessons learned from the human factors side in the Target data breach include:

Management's behavior, including not being transparent and communicating with customers and regulators regarding the cyberattack that eroded customer trust, and the lack of thorough assessment and management of the cybersecurity risks posed by third-party vendors and partners.

Management ensures that organizations do not collect excessive customer data and emphasizes the importance of implementing effective monitoring systems and incident response plans to promptly detect and respond to security incidents.

Last, staff training and risk assessments will be implemented to mitigate the risk of similar incidents.

REFERENCES

1. Jewell M. (2007). "Encryption Faulted in TJX Hacking," NBC News, September 25, 2007. https://www.nbcnews.com/id/wbna20979359.
2. Vijayan J. (2007). "TJX Data Breach: At 45.6M Card Numbers, It's the Biggest Ever," Computer World, March 29, 2007. https://www.computerworld.com/article/2544306/tjx-data-breach--at-45-6m-card-numbers--it-s-the-biggest-ever.html.
3. Espiner T. (2007). "Wi-Fi Hack Caused TK Maxx Security Breach," ZDNET, May 8, 2007. https://www.zdnet.com/article/wi-fi-hack-caused-tk-maxx-security-breach/.
4. Xu W., Grant G., Nguyen H., Dai X. (2008). Security Breach: The Case of TJX Companies, Inc. *Communications of the Association for Information Systems*. Vol. 23, Article 31. https://aisel.aisnet.org/cais/vol23/iss1/31.
5. New Hampshire Department of Justice. (2007, January 17). The TJX Companies, Inc. victimized by computer systems intrusion; provides information to help protect customers. Consumer Protection Bureau. Retrieved September 15, 2021, from https://www.doj.nh.gov/consumer/security-breaches/documents/tjx-20070117.pdf.

6. Jewell M. (2007). "T.J. Maxx Theft Believed Largest Hack Ever," NBC News, March 30, 2007. https://www.nbcnews.com/id/wbna17871485.
7. Kawamoto D. (2007). "TJX Says 45.7 Million Customer Records Were Compromised," CNET, March 30, 2007. https://www.cnet.com/news/privacy/tjx-says-45-7-million-customer-records-were-compromised/.
8. Berg G. G., Freeman M. S., Schneider K. N. (2008). Analyzing the TJ Maxx Data Security Fiasco: Lessons for Auditors. *The CPA Journal*. Vol. 78, No. 8, pp. 34–37. https://www.proquest.com/openview/825912066bc3334bf22efa2fb4bf2a76/1?pq-origsite=gscholar&cbl=41798.
9. Ramasastry Ch., Haggerty N. (2017). "Security Breach at TJX," Ivey Publishing, May 18, 2017. https://hbsp.harvard.edu/product/908E03-PDF-ENG.

Part Eight

Blackmail—Hacktivism

- Ashley Madison
- Dark Web Hosting Provider Hack

25 Ashley Madison

ABSTRACT

This chapter reviews the Ashley Madison case. The breach exposed sensitive information, including the personal details of approximately 32 million users. It analyzes this case from a human factors point of view.

INTRODUCTION

Ashley Madison is a Canadian online dating service and social networking service. The difference between Ashley Madison and other online dating services is that its services are marketed to married people seeking extramarital affairs. The site allows users to hide their account profiles for free. However, users looking to delete their accounts, even those made without the individual's consent, are charged a $19 fee. Ashley Madison claimed to have an international membership of 37.6 million, and all of them assured that their use of this service would be "anonymous" and "100% discreet" [1].

The Ashley Madison cyber incident occurred in 2015. The breach exposed sensitive information, including the personal details of approximately 32 million users, such as their names, email addresses, and sexual preferences.

The data breach severely affected many users, as their personal information was exposed to the public, potentially leading to embarrassment, blackmail, and other negative outcomes. Several lawsuits were filed against Avid Life Media, alleging negligence and breach of privacy, and the company settled for $11.2 million in 2017.

CASE DESCRIPTION

In July 2015, a group calling itself "The Impact Team" stole the user data of Ashley Madison. The group copied personal information about the site's user base and threatened to release users' names and identifying information if Ashley Madison would not shut down immediately.

When the company refused to comply, the hackers followed their threat and released the stolen information online. On August 18 and 20, the group leaked more than 60 GB of company data, including user details [2]. The breach led to widespread public embarrassment and privacy concerns for Ashley Madison users, as well as investigations and lawsuits against the company [3].

A data dump, 9.7 GB in size, was posted to the dark web using an Onion address accessible only through the Tor browser. The files included account details and logins for 32 million social networking site users. Seven years' worth of credit card and other payment transaction details was also part of the dump. AshleyMadison.com claimed to have nearly 40 million users at the time of the breach, all apparently in the market for clandestine hookups [3].

DOI: 10.1201/9781003461739-33

WHAT WE KNOW ABOUT THIS CASE

On July 12, 2015, Avid Life Media employees found a message from the Impact Team threatening to release company and customer data unless the Ashley Madison and Established Men websites were shut down. Attackers gave Avid Life Media a 30-day notice to shut down the sites before they released their information. Following the attackers warning, Avid Life Media announced that an unauthorized party gained access to their systems. After the warning expired, Impact Team published the first major Ashley Madison user data dump on a file-sharing network containing nearly 10 GB of user email addresses.

Later in August 2015, the email addresses disclosed in the first data dump were posted to Pastebin (a website where you can store text online for a set period). Then, a second set of data contained nearly 20 GB of internal data, including Avid Life Media chief executive officer Noel Biderman's emails. The posted data revealed that many governments, military, and corporate addresses were used to sign up for Ashley Madison accounts. Later, krebsonsecurity.com disclosed that numerous Ashley Madison account holders confirmed that their information was published, and the users could search for their email [4–6].

The security experts believe that the breach was facilitated by vulnerabilities in Ashley Madison's security practices, weak passwords, and a lack of encryption for sensitive data.

According to security researcher Gabor Szathmari, the breach was successful due to poor security practices in Ashley Madison's source code. The worst offense was hardcoded security credentials, including database passwords, API secrets, authentication tokens, and SSL private keys. The Ashley Madison website also didn't employ email validation to help screen out bots [7].

Later, a blog post from a cracking group called CynoSure Prime revealed that Ashley Madison failed to use a robust encryption strategy for its user passwords, allowing the group to crack over 11 million passwords in just ten days. The group also published an analysis of the top passwords used by Ashley Madison members, who also exhibited poor password security [4]. In addition, the group published an analysis of the top passwords used by Ashley Madison members. According to the group, "123456" was the most popular password among Ashley Madison users, with over 120k accounts using it to protect them [6]. Later, it was reported that Ashley Madison stored user passwords in plaintext rather than encrypting them [3].

Ashley Madison's sign-up process does not require an email address verification to set up an account, so some site members might have hijacked and used legitimate addresses. For example, one email in the data belonged to former UK Prime Minister (Tony Blair) [3].

A manifesto posted alongside the stolen data—application lifecycle management (ALM). ALM is an integrated system of people, tools, and processes that supervise a software application from initial planning and development through testing and maintenance to decommissioning and retirement. The Impact Team published the ALM, which showed that the customers were charged $19 to erase their profile information. The full Delete netted ALM $1.7 million [8,9].

INFORMATION COMPROMISED

The 2015 Ashley Madison data breach compromised the personal information of its users, including names, addresses, phone numbers, and email addresses. The leaked data also included financial information, such as credit card transactions and billing addresses, and user-generated content, such as messages and chat history. The sensitive nature of the information on the site, which was marketed as a platform for facilitating extramarital affairs, led to widespread media coverage and caused significant privacy concerns for its users in revenue in 2014. The hacking group stated that this claim is false. They explained that users typically pay with credit cards, and their purchase details are not erased as promised. These details include their real names and addresses, which are the most critical pieces of information users expect to be removed [8,10].

WHAT CYBERATTACKERS DO WITH THIS ATTACK?

With the stolen data from the Ashley Madison data breach, cyberattackers can carry out a variety of nefarious deeds. They might, for instance, execute phishing scams using users' email addresses and phone numbers, which could lead to financial loss or identity theft. Additionally, the cyberattackers might use the users' information to extort or blackmail them, which would be very bad for people and their families.

Cyberattackers can engage in various malicious activities with the compromised information from the Ashley Madison case, including blackmail and extortion, identity theft, Phishing, and using the data for other social engineering attacks.

Overall, the data from the Ashley Madison breach were highly sensitive and can be exploited by cyberattackers for various nefarious purposes, causing significant harm to the affected individuals.

HUMAN FACTORS ANALYSIS

The Ashley Madison cyberattack is a particular case from a human factors' standpoint. Several human factors issues can be classified into three categories: management behavior, Ashley Madison IT professionals' behavior, and users' behavior and lack of awareness [10].

Management

Ashly Madison's vulnerabilities were related to management's lack of requirements to secure the application. Ashley Madison's security measures were found to be inadequate, so attackers were able to exploit vulnerabilities in the websites.

Ashley Madison's promise of discretion and privacy for users seeking extramarital affairs. The breach exposed sensitive user data, including personal information and payment details, undermining the trust that users had placed in the platform. Ashley Madison's security infrastructure and practices, including the lack of encryption for sensitive user data and insufficient measures to prevent unauthorized access to user accounts.

Ashley Madison's management behavior after the attack was also problematic since they downplayed the severity of the breach. Their acknowledgment of the incident also came late publicly. The delay in response created uncertainty among users. They also provided limited information about the extent of the data compromise and how it would impact users. The lack of transparency regarding the security vulnerabilities that led to the breach and the steps being taken to address them was another issue.

IT Professional

The Ashley Madison IT professional needed to implement a secure procedure. The application did not have email verification, so users could easily create fake accounts using fabricated email addresses. The lack of email verification also makes it easier for cyberattackers to create accounts using the email addresses of unsuspecting individuals. The lack of email verification increases the risk of account takeover and unauthorized access to sensitive user data. In addition, email verification allows applications to send important notifications, password reset instructions, and security alerts to legitimate users.

Ashley Madison failed to use a robust encryption strategy for its user passwords made it easier for the attackers to obtain and misuse user credentials. It seems that there was insufficient data protection since the attackers were able to penetrate Ashley Madison's systems and databases, indicating that the company may not have had robust measures in place to protect user data from unauthorized access.

The attackers might have employed social engineering techniques to trick employees into divulging sensitive information or providing access to the company's systems.

Users' Behavior and Lack of Awareness

The incident highlighted the importance of exercising caution and discretion when sharing sensitive information online. This case is a perfect example of the need to use strong passwords. The attackers cracked 11 million passwords out of 32 million users. They could not crack the remaining 21 million. The ones they cracked were the weak ones, such as 123456.

CONCLUSION

The Ashley Madison hack served as a warning to the wider technological community and the online dating sector. It showed the flaws of online dating sites and the possible repercussions of not taking cybersecurity seriously. The incident brought to light the significance of considering the human factor in cybersecurity and the possible repercussions of dishonesty and infidelity in intimate relationships.

The impact of the breach was severe, with numerous individuals suffering embarrassment, negative personal and career outcomes, and even suicides. In addition to causing lawsuits against Ashley Madison and its parent company, the data leak also brought to light data breaches' financial and legal repercussions.

The Ashley Madison hack should serve as a reminder of the significance of prioritizing cybersecurity highly across all technological domains, including online dating services. It also emphasizes the necessity of dealing with the human factor in cybersecurity, for instance, through employee training and awareness campaigns. The Ashley Madison hack ultimately warns about the potential repercussions of not taking cybersecurity properly.

REFERENCES

1. Lamoont T. (2016). "Life after the Ashley Madison Affair," The Guardian, February 17, 2016. https://www.theguardian.com/technology/2016/feb/28/what-happened-after-ashley-madison-was-hacked.
2. Iyengar R. (2015). "Hackers Release Data from Cheating Website Ashley Madison Online," Time, August 18, 2015. https://time.com/4002647/ashley-madison-hackers-data-released-impact-team/.
3. Zetter K. (2015). "Hackers Finally Post Stolen Ashley Madison Data," Wired, August 18, 2015. https://www.wired.com/2015/08/happened-hackers-posted-stolen-ashley-madison-data/.
4. Krebs B. (2015). "Online Cheating Site Ashley Madison Hacked," Krebson Security, July 19, 2015. https://krebsonsecurity.com/2015/07/online-cheating-site-ashleymadison-hacked/.
5. Doffman Z. (2019). "Ashley Madison Has Signed 30 Million Cheating Spouses. Again. Has Anything Changed?" Forbes, April 23, 2019. https://www.forbes.com/sites/zakdoffman/2019/08/23/ashley-madison-is-back-with-30-million-cheating-spouses-signed-since-the-hack/?sh=5815640a3878.
6. Nate Lord N. (2017). "A Timeline of the Ashley Madison Hack," Digital Guardian, July 27, 2017. https://digitalguardian.com/blog/timeline-ashley-madison-hack.
7. Szathmari G. (2015). "Credentials in the Ashley Madison Sources," Blog, September 7, 2015. https://blog.gaborszathmari.me/credentials-in-the-ashley-madison-sources/.
8. Constantin L. (2015). "Ashley Madison Coding Blunder Made 11M Passwords Easy to Crack," Computer World, September 10, 2015. https://www.computerworld.com/article/2982959/ashley-madison-coding-blunder-made-11m-passwords-easy-to-crack.html.
9. Brewster Th. (2015). "Ashley Madison Breach Could Expose Privates of 37 Million Cheaters," Forbes, July 20, 2015. https://www.forbes.com/sites/thomasbrewster/2015/07/20/ashley-madison-attack/?sh=4ce906625f48.
10. Basu E. (2015). "Cybersecurity Lessons Learned from the Ashley Madison Hack," Forbes, October 26, 2015. https://www.forbes.com/sites/ericbasu/2015/10/26/cybersecurity-lessons-learned-from-the-ashley-madison-hack/?sh=f5ceab14c82b.

26 Dark Web Hosting Provider Hack

ABSTRACT

This chapter reviews the Dark Web Hosting Provider Hack case. In 2020, a hacker breached Daniel's hosting service, stole the database, and deleted all of the website's databases. Many user's email IDs and passwords were also exposed. This chapter analyzes this case from a human factors point of view.

INTRODUCTION

The sites that people surf daily are only a very small part of the World Wide Web. The Internet is just a method of accessing web pages across the net. This is called the Surface Internet. The Dark Web is a part of the Deep Web known to host various sites where illegal black-market deals are taking place with the shielding of anonymity [1].

The Dark Web is a layer of the internet that is not entered with the usual browsers. Instead, we would need a TOR browser. The TOR stands for "The Onion Router." Onion, as the name implies, means that there are multiple layers before the user is connected to a website. The sites on the Dark Web are usually connected to three stages: Entry-point, TOR Nodes, and Exit Nodes. The Dark Web provides access to everything that should not be accessed by the common people under the current laws. Everything from unregistered weapons, drugs, stolen items, and forged records and credit cards is available for purchase on the Dark Web. Many search engines for the Dark Web can be used, like how we use Google for our daily searches [2].

The Dark Net is mainly used for anonymity and is a haven for whistleblowers, journalists, and activists who, due to stringent laws and punishments, would be unable to send information to the world.

The consumers on the Dark Web are entirely anonymous and cannot be traced. Some popular search engines include DuckDuckGo, Torch, and Ahmia; these sites can be used to find anything, from stolen credit cards to Uranium.

In 2020, a hacker named KingNull breached the hosting service called Daniel's Hosting, stole the database, and deleted all the websites' databases. Many user's email IDs and passwords were also exposed [3].

These attacks caused Daniel's Hosting to close its business, and almost one-third of the domains in the Darknet have been moved to other hosting providers. This was targeted after attackers discovered that this hosting service sheltered child abuse portals.

This chapter investigates the attack in which a hacker leaked a database of the web hosting provider Daniel's Web Hosting.

 DOI: 10.1201/9781003461739-34

CASE DESCRIPTION

The Dark Web is the deep web where around 96% of the internet content is located. Dark Web databases have searching features that users can use to obtain information that is not usually accessible. The Dark Net is a system based on the Deep Web and is concealed by choice. It has been specifically designed to be anonymous.

These can be accessed using the TOR browser or the I2P (Invisible Internet Project), ZeroNet, etc. Onion Routing is done using tunneled encryption. To simplify it, the Tor constructs a Virtual Link between the consumer and each host in the path of the three TOR relays. Every relay then decrypts another layer of encryption to make another relay visible and pass the residual encrypted information. The closing of this relay is where the innermost relays are decrypted, and the underlying initial information is sent over to the user.

The hack was where the hacker KingNull stole the user data and private keys database. The attacker then uploaded these data dumps to a local file-hosting system [3].

WHAT WE KNOW ABOUT THIS CASE

The attacker's KingNull attack was carried out on March 10, 2020. KingNull breached Daniel's hosting portal, stole the data, and then wiped their database clean. Many sensitive data, such as emails, passwords, and consumers' identities, were later leaked onto a local file-sharing system. The leaked data includes 3,671 email addresses, 7,205 account passwords, and 8,580 private keys for ".onion (Dark Web) domains" [3].

It is well known that illegal activities take place in the Deep Web. Alnabulsi et al., in their paper, explain their analysis of the illicit activities that take place in the Deep Web and how they identify them [4]. This hack was done as the web hosting service was also a hideout for child abuse portals as well, among other illicit things.

A firm called Under the Breach was investigated in this case by the threat intelligence firm Under the Breach. According to them, "The leaked database contains sensitive information on the owners and users of several thousand darknet domains."

"This information could substantially help law enforcement track the individuals running or taking part in illegal activities on these darknet sites" [3].

INFORMATION COMPROMISED

Besides all the criminal activities, this leak affected many organizations where anonymity would be the foremost concern. Journalists and whistleblowers who use the Deep Web to relay information to the public despite the cruel regimes they live under were also affected.

This compromise may include undercover spy agencies in the field that use these sites to communicate information to their superiors as well from their undercover jobs.

Since this has been compromised, this could also lead to many imprisonments and even deaths in some countries for information trafficking.

WHAT DID CYBERATTACKERS DO WITH THIS ATTACK?

The hackers could do the following with the data:

- Blackmail the users to release their full identity online; although their emails were released, their identities were not. The attackers could use this to demand financial and other benefits from the users.
- The attackers can also relay the information to the police forces for them to tackle the issues and problems from a law enforcement perspective as well.
- The attackers can also notify hacktivist groups so that they can bring justice to people with cases that law enforcement cannot pursue due to a lack of sufficient information.
- The hackers can use the data to influence certain political and judicial decisions and also use the data as leverage on people.
- The data can be used to bribe officials to do their bidding.
- They can influence the stock market by holding this over the heads of the users and the government as well.
- They can threaten to expose the individuals who are journalists and activists in third-world countries.
- They can use this for their gains and makc transactions for financial benefits.
- They can use this data and private keys to make fake trades on websites in the deep web.
- They can also use this data to bring justice to child abusers, pedophiles, and other attackers. They can give this data to righteous criminals to bring justice to poor children.

HUMAN FACTORS ANALYSIS

The hacker brought down the whole service and ultimately caused it to shut down on March 26, 2020. Many organizations, such as organized crime syndicates and terrorist organizations, have been exposed through this hack, and law enforcement agencies can further pursue the identities of these criminals and incarcerate them for good under the law.

The users and customer database were wiped out so that this could not be returned to life. It appears that the provider did not have robust backup and recovery mechanisms that can help mitigate the impact of breaches and ensure that data can be restored in the event of a security incident.

The providers responsible for secure sensitive Data of users Hosting providers, especially those catering to users on the Dark Web, must have prioritized the security of sensitive data. This includes implementing encryption, access controls, and other measures to protect user data from unauthorized access and exfiltration.

Users of Dark Web hosting services should be educated about the risks associated with operating within the Dark Web ecosystem. This includes practicing good security hygiene, using strong authentication methods, and staying vigilant against potential threats such as phishing attacks.

CONCLUSION

The breach of Daniel's Hosting serves as a reminder of the ever-present cybersecurity threats facing hosting providers and the importance of robust security practices to protect sensitive data and mitigate risks in the Dark Web environment.

The attackers may have a financial gain and hold the data as leverage over the user's heads. Despite the fact that this might be a win for law enforcement agencies that can further pursue the identities of these criminals and incarcerate them for good under the law, it also illustrates that even Dark Web users are not immune to cyberattacks [5].

REFERENCES

1. Ranakoti P., Yadav S., Apurva A., Tomer S., Roy N. R. (2017). Deep web & online anonymity. *2017 International Conference on Computing and Communication Technologies for Smart Nation (IC3TSN),* Greater Noida, India, 2017, pp. 215–219. https://doi.org/10.1109/IC3TSN.2017.8284479.
2. Guccione D. (2021). "What Is the Dark Web? How to Access It and What You'll Find," CSO, July 1, 2021. https://www.csoonline.com/article/3249765/what-is-the-dark-web-how-to-access-it-and-what-youll-find.html.
3. Cimpanu C. (2020). "Hacker Leaks Database of Dark Web Hosting Provider," ZDNET, May 31, 2020. https://www.zdnet.com/article/hacker-leaks-database-of-dark-web-hosting-provider/.
4. Alnabulsi H., Islam R. (2018). Identification of illegal forum activities inside the dark net. *2018 International Conference on Machine Learning and Data Engineering (iCMLDE),* Sydney, Australia, 2018, pp. 22–29. https://doi.org/10.1109/iCMLDE.2018.00015.
5. Winder D. (2020). "Hack Attack Takes Down Dark Web Host: 7,595 Websites Confirmed Deleted," Forbes, March 30, 2020. https://www.forbes.com/sites/daveywinder/2020/03/30/hack-attack-takes-down-dark-web-7595-websites-confirmed-deleted/?sh=209ab98f1435.

Part Nine

Conclusion

- Factors Shaping Vulnerable Behaviors

27 Factors Shaping Vulnerable Behaviors

Abstract

This chapter reviews some fundamentals of human cognition used in cyberattacks. Motivation, perception, attention, memory, decision-making, and errors are analyzed.

INTRODUCTION

The review of several cyberattacks in the previous chapters illustrates how human intention, motivation, actions, and events are the main factors in cyberattacks or data breaches. Various cognitive factors shape vulnerable behaviors. Thus, understating and analyzing human behaviors in cybersecurity requires understanding human cognition and how humans process information.

This chapter reviews some fundamentals of human cognition used in cyberattacks. The main factors are motivation, perception, attention, memory, decision-making, and errors (Figure 27.1).

MOTIVATION

Chapter 3 examined the motive and reasons behind a person's decision to commit a crime. Understanding motivation is crucial for developing effective cybersecurity

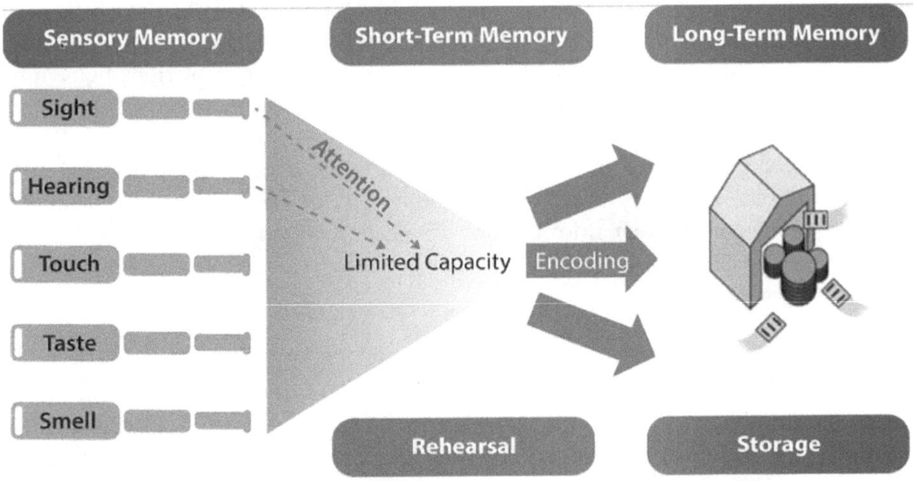

FIGURE 27.1 Some main factors shaping vulnerable behaviors in cybersecurity privacy and trust.

DOI: 10.1201/9781003461739-36

strategies to defend against attackers and ensuring users adhere to security protection rules and policies. It's important to recognize that the motivation behind cyberattacks can be complex and multifaceted, with attackers often driven by a combination of factors.

Understanding the motivation behind cyberattacks helps to define effective cybersecurity strategies, including implementing appropriate security measures, educating employees and users, monitoring suspicious activities, and responding effectively to incidents. Human behavior is highly related to people's motivation; for example, it motivates soldiers to risk their lives for their country. In cybersecurity, threat actors are groups or individuals who might be motivated to cyberattacks for a variety of reasons. Employees who are not motivated may not take security policies seriously or care about the company's assets involuntarily help the attackers to break into the company system. Employees involved in insider threats might be motivated by political ideology, financial gain, fraud, espionage, or a desire for revenge (Chapter 3).

PERCEPTION

Perception of information in cybersecurity is capital for observing, interpreting, and making decisions based on sensory inputs to identify potential security risks and take appropriate actions to mitigate risks.

The main sensory perception in cybersecurity involves the current state of technologies, including vision and audition.

Visual Perception

In human–computer interaction, people who use graphical user interfaces (GUIs) rely heavily on their visual sense to detect potential cybersecurity threats, see notifications, the status of the system, and so on. The user might not perceive a warning when malware tries to install or detect the cure that helps to analyze a phishing mail text. The same might be valid for IT agents monitoring network traffic patterns, reviewing logs and system events, and identifying unusual or suspicious behaviors. A person focusing on the main task on display might not be able to see a target, such as an alert notification [1].

Visual perception is required to recognize patterns, anomalies, and visual cues that may indicate a security breach or malicious activity. Seeing does not mean that people perceive the correct information. Visual perception in cybersecurity often requires that professional agents and users be trained to see relevant information, alert warnings, etc.

Auditory Perception

Sound or auditory signals can also play a role in detecting potential cybersecurity threats, seeing notifications, system status, etc. Many notifications and warnings might be auditive, particularly with all voice user interface technologies. Audio messages might not be heard due to volume, noise in the environment, or just when attention is focused on vision. Auditory signals are particularly important for mobile phones.

For example, security analysts may listen to auditory alerts or notifications from security monitoring systems, such as intrusion detection systems (IDS), firewalls, or security information and event management (SIEM) systems, to detect potential security.

Vulnerable Behavior

Vulnerable behavior refers to patterns of behavior that expose individuals to potential harm, exploitation, or adverse consequences. Analyzing vulnerable behavior helps improve users' awareness and empowers individuals to develop skills and strategies to protect themselves from harm and make informed decisions. It also requires creating environments and systems that prioritize safety, respect, and the well-being of all individuals. Vulnerable behavior might happen at multiple levels.

Signal Not Detected

Signals are generally not detected when the warning messages or notifications are not very salient and stay for a very short period to be seen by the eyes. Sometimes, the signal and warning are shown on the user interface at the location where some users might look or need to be trained to look. For example, many browsers signal that trusted or untrusted site on the browser address bar, but unaware users might not look at that location to see the browser's security warnings. Many browsers have built-in security scanners that warn you before visiting an infected webpage or downloading a malicious file.

Signal Dismissal

Signal dismissal refers to disregarding or ignoring a warning or alert generated by a security system. When systems detect a potential threat or suspicious activity, various cybersecurity systems send alert notifications to ITs monitoring the system and notify them. In this process, a signal dismissal can occur. The reason for dismissal of the signal might happen due to what is called "false positives" when an alert is triggered incorrectly, such as mistaking normal user behavior for malicious activity. If users frequently encounter false positives, they may dismiss or ignore alerts.

Signal dismissal can also result from "alert fatigue." Alert fatigue occurs when an excessive number of alerts desensitizes the responders, causing them to miss, ignore, or delay responses. In cybersecurity, this happens when the security system generates a large volume of alerts, overwhelming human agents. As a result, alert fatigue sets in, leading agents to dismiss alerts without thorough investigation due to the sheer number of notifications they receive [2].

Sometimes, signal dismissal occurs because of lack of awareness. Agents who are not adequately trained in cybersecurity or who do not understand certain alerts' significance may dismiss them unintentionally. They may not recognize the potential risks associated with the alert or be unaware of the consequences of dismissing it [3].

Signal dismissal also happens due to inadequate prioritization of alerts. In this case, the agent received multiple alerts simultaneously and then the agent might prioritize some alerts over others based on their perceived severity or relevance. If an alert is deemed less important or urgent, it may be dismissed or delayed in favor of addressing other higher-priority alerts.

No matter what the cause of signal dismissal is, ignoring security alerts without proper investigation or analysis can leave a system vulnerable to potential threats. Thus, system administrators must carefully evaluate each alert and determine whether further investigation or action is required.

ATTENTION

People focus on relevant stimuli and ignore irrelevant distractors to perform a task. Research suggests different behavioral control mechanisms for elective engagement. For example, load theory suggests two mechanisms of selective attention: a perceptual selection or low perceptual load and a high perceptual load or active mechanism of attentional control that rejects irrelevant distractors even when these are perceived [4–6].

In low perceptual selection, distractors or irrelevant stimuli have a higher chance of capturing attention because spare attentional resources are available. But in high perceptual load, when the task requires a significant amount of attentional resources, those resources are fully utilized by the task at hand, leaving fewer resources available to process irrelevant distractors.

Another theoretical framework of attention that might help to explain the vulnerable behavior related to inattention is the multiple-resource theory (M.R.T.) [7]. M.R.T. postulates that people possess separate fixed-capacity resources for information processing. There are three dimensions in information processing: the processing stage (early vs. late processing), the processing code (spatial vs. verbal information), and the information modality (visual vs. auditory encoding); other sensory channels were not considered in the original version of M.R.T. Thus, based on multiple-resource theory, the concurrent performance of multiple tasks should benefit to the extent that information related to these tasks is presented in different modalities and thus resource competition is reduced.

The brain's selective attention to pertinent information allows people to ignore irrelevant distractions. Research suggests that the ability to ignore irrelevant distractions is not determined just by the intention to be focused or by the separability of the target and distractor stimuli but also by the level and type of processing load involved in the task [6,8–10].

In cybersecurity, attention plays an important role, particularly in threat detection. Human agents (ITs or end-users) need to focus attention on suspicious activities, anomalies, or indicators of compromise within networks, systems, or end-users' behavior. Additionally, human agents pay attention to real-time data, security logs, network traffic, and system behavior to have constant situational awareness, while the threat landscape evolves to make informed decisions.

MULTI-TASKING

People often perform multiple tasks or activities simultaneously or in rapid succession. Thus, multi-tasking requires switching attention and cognitive resources between different tasks to efficiently manage and accomplish various objectives within a given time frame. Multi-tasking can increase productivity and decrease performance errors, particularly when tasks require high levels of attention or concentration.

MEMORY

Human memory is fundamental in information processing. Memory implications on life and how people experience things are essential, from remembering meaningful events to enabling people to execute tasks and achieve goals. It is believed that there are several stages or memory types:

- Sensory memory: It refers to sensory information for very brief periods.
- Short-term memory: It allows a person to recall little bits of information briefly.
- Working memory: It is where data is processed.
- Long-term memory: It refers to information storage of all events, memories, and experiences.

All memory types might play an essential role in human behaviors in cybersecurity. Attackers might use the limitation of sensor memory to target users. For example, a warning might stay briefly on the user's screen, or a temporary password might not be remembered once disregarded from the user's view.

To manage passwords for quickly accessing various online accounts, people use a simple password or keep a written list of passwords. However, as the number of passwords used increases, the incidents of forgetting and mixed-up passwords increase. Information processing is limited to the prioritization of relevant information over irrelevant information. These processing priorities are maintained in working memory so that capacity is allocated with a higher priority to the relevant information. The capacity allocation happens when people engage in an important task where their attention and processing are engaged. For example, they receive a phishing text or mail and click the link without checking the sender's identity or credentials. Task conditions of the higher primary perceptual load reduce the processing of irrelevant distractors simply due to the reduced availability of perceptual resources [6–8,11].

DECISION-MAKING

Cognitive control involves regulating and managing the flow of information in the mind, focusing on selecting and prioritizing relevant information while filtering out what is deemed irrelevant. Thus, processing priorities are actively maintained in working memory so that capacity is allocated with a higher priority to the relevant information. However, if processing the relevant information does not require all available capacity, any remaining capacity is given involuntarily to processing irrelevant information as well (in a simultaneous parallel manner). Consequently, the level of perceptual load in task processing plays a critical role. Task conditions of low perceptual load—for example, detection of a single item or one that pops out from among different things—result in distractor processing even if people attempt to ignore irrelevant distractors. Task conditions of higher perceptual load—for example, an increased number of items or more complex perceptual processing demands like discriminating conjunctions of features—result in reduced processing of irrelevant distractors simply due to reduced availability of perceptual resources [8,11,12]. To better understand people's decision-making approach, approaches such as threat avoidance theory and pattern or style of design making can be applied in cybersecurity.

Threat Avoidance Theory

Threat avoidance theory (TAT) focuses on how people perceive and respond to potential threats agents facing danger. According to this theory, people are motivated to avoid or minimize possible threats to their well-being or survival. In cybersecurity, TAT suggests that human agents engage in threat assessment processes to evaluate the likelihood and severity of a threat and, based on this assessment, determine the most appropriate response. Thus, applying this theory helps in analyzing the factors that influence cybersecurity decision-making and behavior by analyzing the following factors:

- How cyber threats are perceived and assess the potential risks and threats associated with the system.
- How severities of perceived cyber threats are evaluated.
- How the most appropriate responses to mitigate the cyber threat are selected.
- What kind of emotional responses, such as fear, anxiety, or frustration, does the threat create and their influence on decision-making?

Cyber threats can evoke various emotional responses, such as fear, anxiety, or frustration. These emotions can influence decision-making and the willingness to invest time, resources, and effort in cybersecurity measures [13,14].

Pattern or Style of Design Making

Decision-making style refers to the human pattern of making choices and reaching decisions. It is "a habitual pattern used by individuals when making decisions" [15]. Several models and theories classify decision-making styles.

Some studies relate decision-making to demographics and personality traits [15]. Other studies relate decision-making styles to learned response patterns exhibited when confronted with a decision situation.

Scott and Bruce classify decision-making styles into five categories. The rational style is a comprehensive search for information, an inventory of alternatives, and a logical evaluation of other options. The intuitive style focuses more on details in the flow of information rather than the systematic search for and processing of data, and it has a tendency to rely on hunches and feelings. The dependent style searches for advice and guidance from other sources before making decisions—the avoidant style attempts to avoid decision-making whenever possible. Lastly, the spontaneous style considers the immediacy of decision-making and the desire to conclude the decision-making process as quickly as possible [16].

The Vroom-Yetton-Jago Decision Model classifies decision-making into five styles. The first is the autocratic style, in which a single individual without involving others and input from others is not sought or considered. Opposite to the authoritarian style is the consultative style, when an individual seeks input from others before making the final decision. The facilitative style is when the decision-maker involves others in the decision-making process and acts as a facilitator. The delegation style is when the decision-making authority is delegated to others. Finally, the collaborative style is when the decision-making process is highly participative and collaborative [17].

Behavior Pattern Schema

Schemas (or schemata) are units of understanding that can be hierarchically categorized and webbed into complex relationships. Schemas shape through the development of human cognition from childhood to adulthood and through the human lifetime. They are constructs or building blocks that organize our knowledge and experience. The constructs are all interconnected and constitute a complex interactive network. A schema guides information acceptance and retrieval: it affects how humans process new information and retrieve old information from long-term memory. Piaget and Cook called schema as the core building block [18,19].

For example, think of a restaurant; you probably get an immediate mental image of the restaurant, the process of being seated, offering the menu, taking people's orders, and the payment process. However, this mental image of the restaurant might be different from country to country. Now, imagine the questions you might ask if you want to recall the process and questions you would answer when you call your bank for a service.

Once a schema is repeated over and over, it becomes expected; if, by any chance, one step was to be added or changed, that would interrupt people's expectations and confuse them. For example, when you call to reserve a table at a restaurant on the phone, you are used to giving your name and phone number. Suppose the reservationist were to also ask for your driver's license number, that would violate your known schema for this activity and appear very odd to you. Unless a specific, convincing reason were to be given, you would very likely refuse to answer the question. During our lives, as we learn and discover the world around us, our schemas expand and get complex. The more we know, the more significant and complex our schemas become. However, the more we are aware, the easier it is to remember new information related to the schema. Thus, since the data exists in our heads, we can relate to it and organize and predict our actions. Activity schemas are called scripts. We all have many scripts in our long-term memory associated with the various activities we have engaged in throughout our lives.

Social engineers in the digital world use schema concepts. They carefully prepare and investigate the process, establish an accurate script of each situation, and then find believable responses that capture their victims' trust [20].

REASONING

Reasoning (cognitive processes and logical thinking) is fundamental in decision-making when facing security challenges or breaches. There are overall four main types of reasoning: deductive reasoning, inductive reasoning, abductive reasoning, and reasoning by analogy [21].

Deductive Reasoning

Deductive reasoning is a logical process of concluding by premises to specific situations or cases. It involves moving from general statements, premises, to specific conclusions.

For example:

Premise 1: All mammals are warm-blooded animals.
Premise 2: A dog is a mammal.

Conclusion: Therefore, the dog is a warm-blooded animal.

In cybersecurity, deductive reasoning helps to consider a cybersecurity incident where there's a suspicion of access request; for example, consider the following scenario when files are received from an external source in an organization:

Premise 1: All files received from external sources should be scanned for malware.
Premise 2: File "A" is received from an external email.
Conclusion: Therefore, File A should be scanned for malware. Based on the premises, File A is received from an external source. If the scan identifies malware, appropriate actions such as quarantining the file or notifying the user may be taken.

Inductive Reasoning

Inductive reasoning involves drawing general conclusions from specific observations or evidence. It is commonly used in everyday reasoning, scientific research, and data analysis.

It is a bottom-up approach in which specific instances or examples are used to infer a broader pattern or principle. In cybersecurity, professionals use logical reasoning to analyze the situation, identify the root causes, and develop effective solutions. In cybersecurity, inductive reasoning helps to consider understanding the nature of a security incident. For example, consider this scenario:

Evidence 1: Multiple employees report similar phishing emails.
Evidence 2: All the characteristics of these emails are identified in these phishing emails.
Conclusion: Most probably, this is a phishing campaign targeting the organization, using common tactics and characteristics across multiple incidents. Thus, inductive reasoning is applied by drawing a general conclusion about

the existence of a broader phishing campaign based on specific instances and shared characteristics observed in those instances. Inductive reasoning in cybersecurity helps organizations make informed decisions and take proactive measures to address potential threats based on patterns and trends identified through specific observations and incidents.

Abductive Reasoning

Abductive reasoning is logical reasoning. It is used to make an educated guess or hypothesis to explain a particular observation, phenomenon, or set of facts. Abductive reasoning is used when people seek the most plausible explanation based on available evidence. For example, consider this scenario: In an organization, the cybersecurity agent observes the following.

Observation 1: The network monitoring system detects a significant increase in outgoing network traffic from a specific workstation.
Observation 2: The affected workstation is running a legitimate software application.
Observation 3: There are no reports of system vulnerabilities or malware infections in the organization.
Conclusion: This increase in outgoing network traffic might be due to the legitimate software application on the workstation, possibly due to a misconfiguration or a recent update. The agent might also formulate other hypotheses based on the factual observation. In this case, the cybersecurity team may investigate the specific software application on the workstation, looking for recent changes, updates, or misconfigurations that could explain the unusual network behavior. The agent also might consider other variables or information to check the hypothesis and may also consider the possibility of a compromised version of the software.

In this example, abductive reasoning is applied to generate the best hypothesis that explains the observed evidence. It involves considering multiple factors and forming a conclusion that is both plausible and fits the available information. Abductive reasoning in cybersecurity is valuable for identifying potential threats or incidents and guiding further investigation to address the underlying causes.

Analogy Reasoning

Analogy reasoning is used when drawing parallels or similarities between two or more different situations, concepts, or objects. In this type of reasoning, people rely on the assumption that if two things share certain similarities, they are likely to share other characteristics. Thus, analogy is considered a top-down cognitive process. In analogy reasoning, people use pre-existing knowledge, context, or higher-level cognitive functions to interpret and understand information. For example, consider this scenario: In an organization, the cybersecurity agent observes the following.

Analogy: "Virus detection can be understood by considering the flu or flu-like viruses. If a virus is going around and making people sick, a city might protect its people by testing people for all types of known viruses at the city's entry points. If a vaccine exists for the type of infection, people may be required to vaccinate before entering the city. Otherwise, they may be required to quarantine until the risk of viral infection has passed. Virus detection applications automatically do this for computer viruses" [22].

In this example, analogy reasoning helps convey cybersecurity concepts by leveraging a familiar scenario to explain. Analogies can be powerful tools for communication and comprehension in cybersecurity education and awareness.

Using effective reasoning helps users and professionals to analyze the situation, identify the issues, and develop effective solutions. Effective reasoning is key to cybersecurity behavior in risk assessment, forensic analysis, security policy development, etc.

ERROR

Human errors often contribute to the success of cyberattacks. Cyberattackers exploit vulnerabilities that arise from peoples' errors due to mistakes, slips, lapses in judgment, or false reasoning.

We are all occasionally making errors during our daily activities. Some errors might have minor consequences, and some more severe. We are frequently reminded of the famous statements: "To Err is Human" [22] or "If an error is possible, someone will make it" [23]. Not surprisingly, people need to correct their interactions with computers. According to Sicur 2024 report, 82% of data breaches are related to human error [24].

Aware of users' tendency to make such errors, attackers exploit these opportunities when designing their attacks. Despite increased awareness about avoiding unknown links, many people still click on them. Phishing remains the easiest method for attackers to obtain people's credentials.

Two scientists have focused their work on human error: Donald Norman and James Reason.

Donald Norman [25,26] has classified human error in the computer system into two categories: mistakes and slips. A mistake is considered an error in the intention. They are decision failures. Mistakes are divided into rule-based mistakes and knowledge-based mistakes. Mistakes are referred to as errors of commission. They happen when people do the wrong action when they believe it is correct and when users do many tasks at once or run out of time. Thus, people with a false understanding of how the system works act intentionally. Still, their actions do not result in how they are expected because they need to understand how a specific system functions. Making mistakes is a core part of the human experience.

Slips (automatic behavior) and lapses (skill-based errors) are considered errors in execution. In this case, the user knows the right action but did it wrong. Slips and lapses occur in familiar tasks performed without conscious attention. Slips are also errors of commission; for example, an operator moves the switch upward instead of downward. Slips can be classified into small classes based on the mechanisms that are the most likely causes. For example, after clicking a phishing link, some people immediately realize they have made an error, but that might be too late. These errors might be performing the correct action on the wrong object, which will trigger an incorrect movement, or an error by association that will trigger a false action.

Lapses, however, are errors of omission (not doing something that should be done)—short-term memory loss results in errors when the operator omits procedures or matters necessary to perform essential tasks.

Reason [27] classified different actions that lead to an error into two groups: unintended action (slip and lapse) and intended action (mistake and violation). The intended action error results from conscious decision-making or subconscious processing.

James Reason developed a widely used error classification system known as the "Swiss Cheese Model" or the "Reason Model." This model categorizes errors based on their underlying causes and the layers of defense in a system. Here are the critical error types in James Reason's classification:

1. Active Errors: Active errors refer to the immediate actions or decisions made by individuals that directly lead to an error or an adverse event. These errors are often the most visible and occur at the frontline level. Examples of active errors include slips, lapses, mistakes, and violations made by operators or workers in real time.
2. Latent Errors: Latent errors are underlying system flaws or conditions that may not have an immediate impact but can influence the occurrence of active errors. These errors are not immediately visible and can exist in the system's design, organization, processes, or maintenance. Latent errors can remain dormant for a long time until triggered by certain conditions. They are typically introduced during system development or changes made to the system. Examples of latent errors include inadequate training, poor communication channels, unclear procedures, or equipment design flaws.
3. Preconditions for Errors: Preconditions for errors are factors that set the stage for the occurrence of errors. These factors create an environment where errors are more likely to happen. They can be related to individual factors, team dynamics, organizational culture, or environmental conditions. Examples include time pressure, fatigue, inadequate resources, insufficient supervision, poor teamwork, or lack of feedback mechanisms.
4. Unsafe Acts: Unsafe acts refer to intentional or unintentional actions by individuals that deviate from established rules, procedures, or standards. These acts can increase the likelihood of errors or lead to adverse events. Unsafe acts can include violating safety protocols, shortcuts, bypassing safeguards, or failing to follow established procedures.

Human errors are the enablers of many cyberattacks. The causes of the errors are different [27]. They may be related to forgetting, inattention, and wrong decision-making strategies. The gravity and consequences of errors differ depending on the human agent's role. For example, an Information Technology professional who forgets to install updated security patches leaves the system dangerously vulnerable, and the consequences of that error might be disastrous. The individuals who click a phishing email might be susceptible to data loss, ransomware attacks, or identity theft. The attackers might use common errors people generally make to trick them into revealing their credentials. Users might plug infected USB drives or devices into computers, introducing malware into the network. People who are unaware of the danger of weak passwords use weak or easily guessable passwords or use the same password across multiple accounts, making it easier for attackers to gain unauthorized access.

CONCLUSION

Human vulnerable behaviors in cybersecurity include a range of behaviors and patterns that make individuals vulnerable to various cyberattacks. Human behaviors are complex and involve many psychological, cognitive, social, environmental, and biological factors.

As mentioned before, some of the vulnerable behaviors used in cyberattacks are directly related to human cognition. These include perception, attention, memory, decision-making, and error. For instance, not seeing or ignoring a notification alert, forgetting to update security patches, or mistakenly clicking on a phishing link.

Some vulnerable behaviors are associated with emotions. In an organization where systemic inequalities and discrimination exist, this might create an angry emotion among some employees, leading to a sentiment of revenge and resulting in an insider threat. The potential consequences of such behavior can be severe, as employees might then be very imaginative when conducting inside threats.

Some vulnerable behaviors have their root in social proofs, cultural norms, and societal expectations. For example, in a company where security policies are not taken seriously and many employees do not follow the rules, this lax attitude might put the system at a high risk of vulnerability.

In summary, understanding human vulnerable behavior in cybersecurity is crucial. It helps people become aware of risks and modify susceptible behavior by adopting defensive strategies.

REFERENCES

1. Wickens Ch. (2021). Attention: Theory, Principles, Models, and Applications. *International Journal of Human-Computer Interaction*. Vol 37, No. 5, pp. 403–417. https://doi.org/10.1080/10447318.2021.1874741.
2. Reeves A., Delfabbro P., Calic D. (2021). Encouraging Employee Engagement with Cybersecurity: How to Tackle Cyber Fatigue. *SAGE Open*. Vol. 11, No. 1, p. 21582440211000049. https://journals.sagepub.com/doi/full/10.1177/21582440211000049.
3. Jeong R. (2019). *Children and Adults' Perception of Signal Colours, Symbols, and Words in the Context of Cybersecurity Warnings*, Master of Arts: M.A., Carleton University. https://curve.carleton.ca/29261a25-4bcd-4bcb-82a4-205a5d6fdaaa.
4. Zeng M. (2018). *Dynamic Cybersecurity Warnings: Subliminal and Supraliminal Approaches*, Dissertation for the Degree the Doctor of Philosophy, The University of Alabama in HuntsvilleProQuest Dissertations Publishing. https://www.proquest.com/docview/2282209422/previewPDF/3BCAE41BD6BA4C39PQ/1?accountid=10361.

5. Andersen E., Goucher-Lambert K., Cagan J., Maier A. (2021). Attention Affordances: Applying Attention Theory to the Design of Complex Visual Interfaces. *Journal of Experimental Psychology*, Applied. Advance online publication. https://doi.org/10.1037/xap0000349.
6. Lavie N., Hirst A., de Fockert J. W., Viding E. (2004). Load Theory of Selective Attention and Cognitive Control. *Journal of Experimental Psychology General.* Vol. 133, No. 3, pp. 339–54. https://pubmed.ncbi.nlm.nih.gov/25085738/.
7. Lavie N. (1995). Perceptual Load as a Necessary Condition for Selective Attention. *Journal of Experimental Psychology: Human Perception and Performance*, Vol. 21, pp. 451–468.
8. Lavie N., Tsal Y. (1994). Perceptual Load as a Major Determinant of the Locus of Selection in Visual Attention. *Perception & Psychophysics.* Vol. 56, No. 2, pp. 183–197. https://doi.org/10.3758/bf03213897.
9. Wickens CD. (1984). Processing resources in attention. In R. Parasuraman & R. Davies (Eds.), *Varieties of Attention.* New York Academic Press, New York, NY, pp. 63–110.
10. Kramer A. F., Wiegmann D. A., Kirlik A. (2006). *Attention: From Theory to Practice*, Oxford University Press, 288 pages.
11. Konstantinou N., Beal E., King J. R. et al. (1985). Working Memory Load and Distraction: Dissociable Effects of Visual Maintenance and Cognitive Control. *Attention, Perception, & Psychophysics.* Vol. 76, pp. 1985–1997. https://pubmed.ncbi.nlm.nih.gov/25085738/.
12. Camina E.et al. (2017). The Neuroanatomical, Neurophysiological and Psychological Basis of Memory: Current Models and Their Origins. *Frontiers in Pharmacology.* Vol. 8, p. 438. https://doi.org/10.3389/fphar.2017.00438.
13. Singh T., Johnston A. C., D'Arcy J., Harms P. D. (2023). Stress in the Cybersecurity Profession: A Systematic Review of Related Literature and Opportunities for Future Research. *Organizational Cybersecurity Journal: Practice, Process and People.* Vol. 3, No. 2, pp. 100–126. https://www.researchgate.net/publication/368839277_Stress_in_the_cybersecurity_profession_a_systematic_review_of_related_literature_and_opportunities_for_future_research.
14. McClelland R. T. (2010). The Pleasures of Revenge. *The Journal of Mind and Behavior.* Vol. 31, No. 3/4, pp. 195–235.
15. Scott S. G., Bruce R. A. (1995). Decision-Making Style: The Development and Assessment of a New Measure. *Educational and Psychological Measurement.* Vol. 55, No. 5, pp. 818–831.
16. Gratian M., Bandi S. et al. (2018). Correlating Human Traits and Cyber Security Behavior Intentions. *Computers & Security.* Vol. 73, pp. 345–335. https://www.sciencedirect.com/science/article/abs/pii/S0167404817302523.
17. Vignesh M. (2020). Decision Making Using Vroom-Yetton-Jago Model with a Practical Application. *International Journal for Research in Applied Science & Engineering Technology.* Vol. 8. https://doi.org/10.22214/ijraset.2020.31876?
18. Moallem A. (2019). Chapter 7 - Social engineering. In A. Moallem (Ed.), *Human-Computer Interaction and Cybersecurity Handbook.* Boca Raton, London, New York: CRC Press, pp. 139–156.
19. Piaget J., Cook M. T. (1952). *The Origins of Intelligence in Children*, New York: International University Press.
20. Abroshan H., et al. (2021). Phishing Happens beyond Technology: The Effects of Human Behaviors and Demographics on Each Step of a Phishing Process. *IEEE Access.* Vol. 9, pp. 44928–44949. https://doi.org/10.1109/ACCESS.2021.3066383.
21. Reid D.A., Knipping Ch. (2010). Types of reasoning. In *Proof in Mathematics Education.* Boston, MA: Drill Publisher, pp. 83–128. https://doi.org/10.1163/9789460912467_009.

22. Moallem A. (2022). Chapter 4. In *Understanding Cybersecurity Technologies: A Guide to Selecting the Right Cybersecurity Tools*. CRC Press/Routledge, p. 50.
23. Croskerry P. (2010). To Err Is Human – and Let's Not Forget It. *CMAJ*. Vol. 182, No. 5, p. 524. https://www.ncbi.nlm.nih.gov/pmc/articles/PMC2842843/#:~:text=Alexander%20Pope%2C%20poet%20of%20the,safety%3A%20To%20Err%20is%20Human.
24. IFEMA MADRID. (2024). "Sicur Cyber 2024: 82 Percent of Data Breaches Are Related to Human Error," git-sicherheit, August 2024. https://git-sicherheit.de/en/news/sicur-cyber-2024-82-percent-of-data-breaches-are-related-to-human-error/.
25. Norman D. (1988). Knowing What to Do: Constraints. *The Psychology of Everyday Things*, New York: Basic Books, p. 140.
26. Norman D. (1993). Design Rules Based on Analyses of Human Error. *Communications of the ACM*. Vol. 26, pp. 254–258. https://doi.org/10.1145/2163.358092.
27. Reason J. (1990). *Human Error*, Cambridge, England: Cambridge University Press, 302 pages.

Index

Note: **Bold** page numbers refer to tables and those in *italics* refer to figures.

MIX
Papier | Fördert gute Waldnutzung
FSC® C083411

Kolibri 360 GmbH
Ferdinand-Jühlke-Straße 7
99095 Erfurt, Deutschland
produktsicherheit@kolibri360.de